Advances in Planetary Science – Vol. 3

PLANETARY HABITABILITY AND STELLAR ACTIVITY

Advances in Planetary Science

Print ISSN: 2529-8054
Online ISSN: 2529-8062

Series Editor: Wing-Huen Ip *(National Central University, Taiwan)*

The series on Advances in Planetary Science aims to provide readers with overviews on many exciting developments in planetary research and related studies of exoplanets and their habitability. Besides a running account of the most up-to-date research results, coverage will also be given to descriptions of milestones in space exploration in the recent past by leading experts in the field.

Published

Vol. 3   *Planetary Habitability and Stellar Activity*
by Arnold Hanslmeier

Vol. 2   *Origin and Evolution of Comets:*
Ten Years after the Nice Model and One Year after Rosetta
by Hans Rickman

Vol. 1   *Nuclear Planetary Science: Planetary Science Based on Gamma-Ray,*
Neutron and X-Ray Spectroscopy
by Nobuyuki Hasebe, Kyeong Ja Kim, Eido Shibamura and
Kunitomo Sakurai

Advances in Planetary Science – Vol. 3

PLANETARY HABITABILITY AND STELLAR ACTIVITY

Arnold Hanslmeier
University of Graz, Austria

World Scientific

NEW JERSEY · LONDON · SINGAPORE · BEIJING · SHANGHAI · HONG KONG · TAIPEI · CHENNAI · TOKYO

Published by

World Scientific Publishing Co. Pte. Ltd.

5 Toh Tuck Link, Singapore 596224

USA office: 27 Warren Street, Suite 401-402, Hackensack, NJ 07601

UK office: 57 Shelton Street, Covent Garden, London WC2H 9HE

Library of Congress Cataloging-in-Publication Data
Names: Hanslmeier, Arnold, author.
Title: Planetary habitability and stellar activity / Arnold Hanslmeier (University of Graz, Austria).
Other titles: Advances in planetary science ; v. 3.
Description: Singapore ; Hackensack, NJ : World Scientific, [2018] | Series:
    Advances in planetary science ; vol. 3 | Includes bibliographical references and index.
Identifiers: LCCN 2018002459| ISBN 9789813237421 (hardcover ; alk. paper) |
    ISBN 9813237422 (hardcover ; alk. paper)
Subjects: LCSH: Extrasolar planets. | Habitable planets. | Stellar activity.
Classification: LCC QB820 .H363 2018 | DDC 523.2/4--dc23
LC record available at https://lccn.loc.gov/2018002459

British Library Cataloguing-in-Publication Data
A catalogue record for this book is available from the British Library.

For any available supplementary material, please visit
http://www.worldscientific.com/worldscibooks/10.1142/10916#t=suppl

Desk Editor: Ng Kah Fee

Typeset by Stallion Press
Email: enquiries@stallionpress.com

Preface

The first exoplanets were detected more than two decades ago. Since then, our knowledge about exoplanets has increased exponentially. Several thousands of these objects have been identified, and many parameters such as their sizes, masses, temperatures, distances to their host stars etc. have been confirmed. There are several reasons for this rapidly growing field of astrophysics. On the one hand, exoplanets can be detected with relatively small telescopes, only a precise photometry is required to detect their transits; on the other hand, several space missions have been devoted to increasing our knowledge about these strange worlds. However, there is one open question: Could life exist elsewhere in the universe?

In this book we mainly deal with the influence that host stars could have on exoplanets. In the first chapter, we give an overview of the objects in our solar system. To understand the complex interaction between stellar activity and planets, it is extremely useful to study these interactions on the different planets and objects in the solar system. Some planets, like Earth or Jupiter, have magnetic fields that provide a shielding against charged energetic particles coming mainly from our host star the Sun. Others, like Mercury, only have weak protection and are very close to the Sun.

In the second chapter, we give an overview on the Sun and its activity. Since the Sun is the only star where this activity can be studied and observed in detail, it is a proxy for stellar activity in general. Only in the case of the Sun can we observe processes during which huge amounts of energy are released (flares, CMEs) within a short time span and the propagation of these particles and radiation to the planets can be measured almost *in situ* by space missions. The activity of the Sun is periodic, but intermittent phases also exist during which solar activity is almost absent. We discuss all these aspects of space weather.

In the third chapter, exoplanets are discussed. We briefly describe how exoplanets can be found and what types of exoplanets have been detected so far. Still, our detection methods are limited, and therefore the observed distribution of exoplanets is strongly biased because only large objects in the vicinity of their host stars can be detected easily.

Stellar activity on a solar level is extremely hard to detect but, by comparing our Sun with other stars, we can understand how solar activity might have been in the past and evolve in the future. In the fourth chapter, stellar activity is discussed and an overview on the basic principles of stellar structure and evolution is also provided.

The main topic of the book is habitability and stellar activity. Several aspects of habitability are presented in Chapter 5. Since we know of only one sample of life in the universe, life on Earth, we give some arguments that life on exoplanets could evolve under conditions similar to the conditions on Earth and some solar system objects. Therefore, a habitable zone around a star can be defined as the region where water on hypothetical planets could exist in liquid form. Water seems to be the basic element for life. We will describe where such a habitable zone can be expected (a) around a star and (b) around a giant planet. A galactic habitable zone also exists. Too close to the center of a galaxy and also too far life seems to be impossible. This is not a real strong limitation to the possible number of habitable planets around stars since a galaxy contains several 100 billions of stars and, even in the worst case, if we are the only habitable planet in our galaxy, billions of galaxies exist in the universe. So therefore, billions of habitable worlds could exist.

In the last chapter, we again address to the question of the influence of stellar activity on planetary habitability. We will discuss in detail how the location of habitable zones around a star changes during stellar evolution. Within the next 4 billion years, the Sun will progressively evolve into a red giant and the Earth will become part of the Sun making it an extremely hostile hot environment for life. What would happen if a superflare on the Sun occurs? Could this be a threat to our highly evolved technological society? To answer such questions it is again useful to compare our Sun with other stars that are younger or older than the Sun.

Each chapter of this books gives some introductory material for the interested reader and then actual literature is cited on the several topics.

Using this literature, the reader can penetrate deeper into the actual research topics of this very rapidly evolving field.

The author is very grateful to several colleagues who went through the manuscript and gave valuable suggestions and hints: Mag. Ines Juvan, Dr. Peter Leitner, Dr. Martin Leitzinger, Mag. Isabell Piantschitsch and Mrs. W. Isaacs for English editing.

A. Hanslmeier

Contents

List of Figures

List of Tables

Chapter 1

The Solar System

In this chapter we review the basic properties of the solar system, with emphasis on the planetary bodies. The central body of the solar system, the Sun, will be described in a separate chapter. Detailed information about planets can be only obtained for solar system planets, and therefore they serve as proxy planets for all other known exoplanets.[1] Also the interaction with stellar activity can be studied in detail in the case of the solar system.

1.1 Objects of the solar system: An overview

The solar system consists of the Sun as the central star, 8 planets, dwarf planets, asteroids, comets and other small solar system bodies and particles. Whereas the main mass of the solar system is comprised inside the Sun (about 99.8%), the main angular momentum is distributed over the orbiting planets. The orbits of the planets are almost coplanar. These facts provide some important hints for the solar system formation.

1.1.1 *The planets*

There are three groups of planets in the solar system:

- **Gas giants:** Jupiter (radius about 70,000 km, its mass exceeds twice that of all other known planets) and Saturn (radius about 60,000 km); both giant planets possess a rocky core with a mass of about 10 Earth masses. In Fig. 1.1 a comparison between the size of Earth and Jupiter is shown. Note the aligned clouds of Jupiter's atmosphere parallel to its equator. Jupiter and Saturn are mostly composed of hydrogen and helium.

---

[1]See the link *exoplanets.eu*, a full catalogue of detected and confirmed exoplanets.

Fig. 1.1   Comparison of Earth and Jupiter. Credit: NASA.

- **Ice giants:** Uranus and Neptune; these planets contain large amounts of water and ammonia, NH_3, methane, CH_4 and silicates/rock. The radii of Uranus and Neptune are about 25,000 km.
- **Terrestrial planets:** Mercury, Venus, Earth and Mars. Earth and Venus have radii larger than 6,000 km. Mars has a radius of approximately 3,500 km and Mercury of about 2,500 km. All terrestrial planets have an atmosphere. The atmosphere of Mercury is extremely thin in comparison to the atmosphere of Venus which is extremely thick. In the case of Venus, we can study the interaction of solar energetic particles (SEPs) with it's atmosphere since Venus does not have a magnetic field. We on Earth are shielded from charged solar particles by a magnetic field.

Some basic planetary data are listed in Table 1.1. Distances in the solar system are given in Astronomical Units (AU).

1 AU denotes the mean Sun–Earth distance $= 150 \times 10^6$ km.

Mass units are often given in Earth masses ($M_\oplus$), and the radius is also given in units of the Earth's radius ($R_\oplus$). For larger exoplanets, units are given in units of Jupiter's mass (M_J) or Jupiter's radius R_J.

Table 1.1. Some important parameters of the planets in the solar system. D denotes the distance from the Sun, P the period of revolution about the Sun, R the planetary radius and P_{Rot} the rotation period of the planet.

Planet	$D \times 10^6$ km	P	$2R$ (km)	P_{Rot}
Mercury	57.91	88.0 d	4879 (0.38)	58.65 d
Venus	108.21	224.7 d	12100 (0.94)	−243.02 d
Earth	149.6	365.25 d	12742 (1.0)	23 h 56 m
Mars	227.92	687.0 d	6780 (0.53)	24 h 37 m
Jupiter	778.57	11.75 y	139822 (10.97)	9 h 55 m
Saturn	1433.53	29.5 y	116464 (9.14)	10 h 40 m
Uranus	2872.46	84 y	50724 (3.98)	−17.24 h
Neptune	4495.06	165 y	49248 (3.87)	16.11 h

1.1.2 Moons

Except for Venus and Mercury all planets have moons. Jupiter's moon Ganymede and Saturn's moon Titan are slightly larger than planet Mercury. In Fig. 1.2 some selected large moons and Earth are shown for comparison.

The four largest satellites of Jupiter, namely Io, Europa, Ganymede and Callisto, are also called Galilean satellites, since Galileo Galilei first detected them in 1609 with his telescope. Some of these moons are especially interesting objects for astrobiologists since there is evidence for the existence of a liquid ocean below an icy crust (Jupiter's moon Europe for example, see Singer *et al.* [2009]). The icy surface of Europa is shown in Fig. 1.3. Such a subsurface ocean remains liquid because of tidal heating: the planet's satellite becomes continuously deformed and stretched by the tidal forces of its nearby giant planet. This is illustrated in Fig. 1.4. The strongest effect of tidal heating can be seen on Jupiter's satellite Io. The surface of Io is covered by several active volcanoes. Io is about the same distance from Jupiter as the Moon is from the Earth; however, Io experiences much stronger tidal stretching because Jupiter is over 300 times more massive than the Earth. Io's rocky surface bulges up and down by as much as 100 m!

One of the first papers dealing with tidal heating and a subsurface ocean of Europa was Reynolds *et al.* [1987].

Tidal heating leads to the concept of a habitable zone around a giant planet (see Chapter 5 about habitable zones). A satellite in such a tidally heated habitable zone could be a possible candidate for life.

Io's orbit is kept from being exactly circular due to the gravitational influence of its Galilean neighbor Europa and the more distant Ganymede.

Fig. 1.2   Selected moons of planets and the Earth for comparison. Credit: NASA.

Fig. 1.3   Icy surface of Jupiter's moon Europa. Credit: NASA.

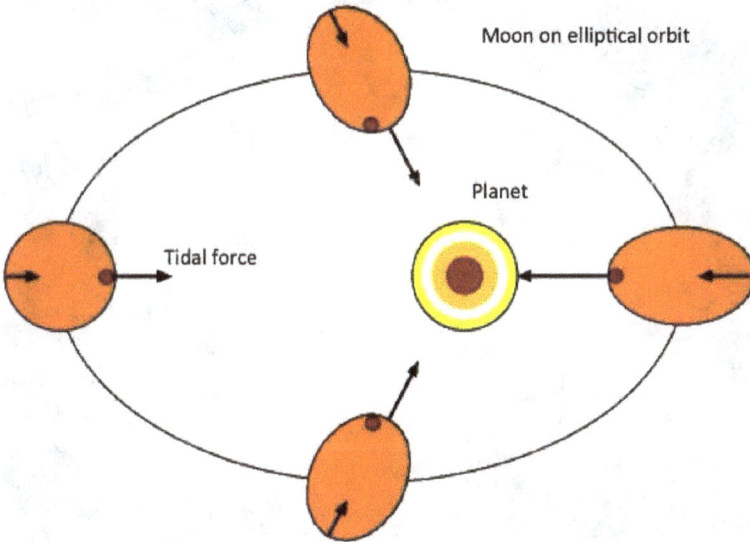

Fig. 1.4   Tidal heating. A satellite on an elliptical orbit is heated by tidal forces.

Io, Europa and Ganymede are in 4:2:1 orbital resonance that keeps their orbits elliptical. This means that for every four orbits of Io, Europa orbits twice and Ganymede orbits once.

In Barnes *et al.* [2013] so-called tidal Venuses are discussed. These are hypothetical terrestrial planets orbiting low-mass stars. They are tidally heated, and this heating induces a runaway greenhouse effect. If this effect is of long duration, all of the hydrogen can escape, and therefore no longer does water exists on such a planet. Without water, there will be no life.

1.1.3   *Dwarf planets*

Pluto (Fig. 1.5) was discovered by Clyde Tombaugh in 1930, and was originally considered the ninth planet from the Sun. Its planet status was questioned following the discovery of several objects of similar size in the Kuiper Belt in 1992. In 2005, Eris, which is 27% more massive than Pluto, was discovered, which led the International Astronomical Union (IAU) to define the term "planet" formally for the first time the following year. Pluto was reclassified by the IAU as a member of the new "dwarf planet" category.

Fig. 1.5   Full disk image of Pluto. This image was taken from a distance of 450,000 km. Credit: NASA/New Horizons.

The official definition of a dwarf planet is as follows:

A dwarf planet is a planetary-mass object that is neither a planet nor a natural satellite. That is, it is in direct orbit of the Sun, and is massive enough for its gravity to crush it into a hydrostatic equilibrium shape (usually a spheroid), but has not cleared the neighborhood of other material around its orbit.

1.1.4 *Asteroids and belts*

The first discovered object in the asteroid belt, Ceres, is now classified as a dwarf planet. Asteroids are minor planets, and most of them are irregularly shaped. A comparison of several objects in size is shown in Fig. 1.6.

Between the orbits of Mars and Jupiter there is the main Asteroid Belt containing several 10^5 objects. Only a few of these asteroids have diameters larger than 100 km. Beyond the orbit of Neptune, there is the Kuiper Belt containing also several 10^8–10^{10} objects. Some recent studies of the Kuiper Belt are given e.g. by Shannon *et al.* [2016]. In its youth, the Kuiper Belt

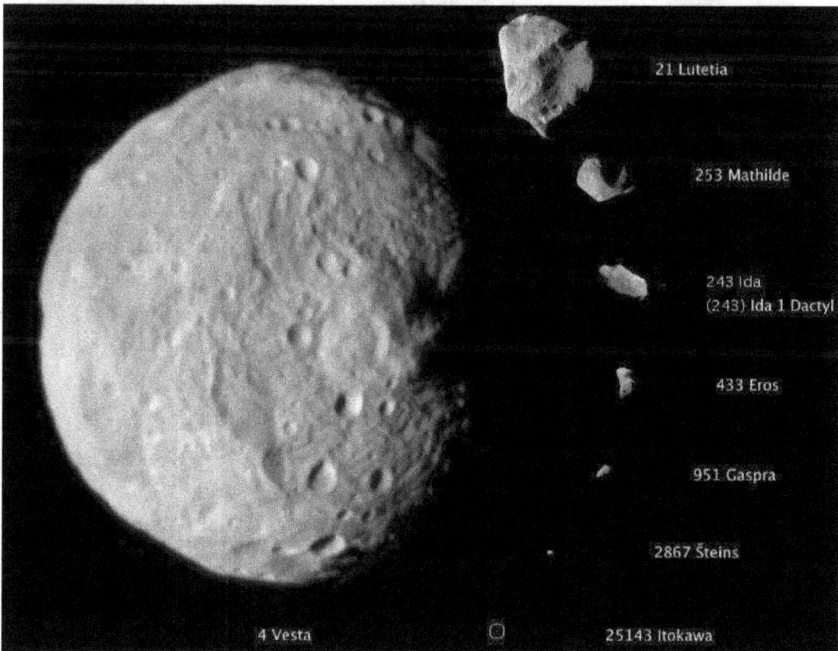

Fig. 1.6 Comparison of sizes of several asteroids. The mean diameter of Vesta is about 516 km. Credit: NASA.

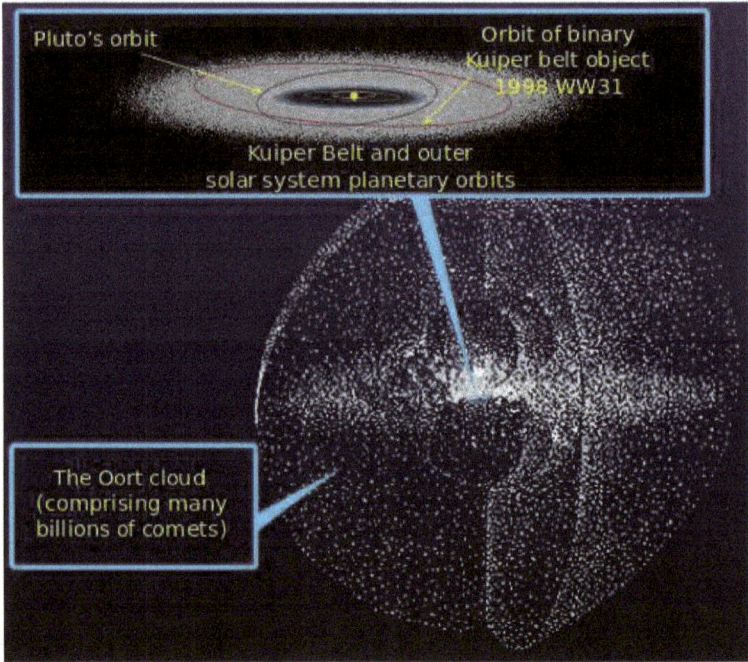

Fig. 1.7   The outer solar system containing the Kuiper Belt and the Oort cloud. Credit: NASA, William Crochot. Available at: http://herschel.jpl.nasa.gov/solarSystem.shtml.

may have compared to the dust rings observed now around such stars as GG Tau and HR 4796A (Luu and Jewitt [2002]). All the objects discussed so far orbit around the Sun in nearly the same plane. The Oort cloud contains even more objects than the Kuiper Belt, and these objects are distributed over a spherical region at heliocentric distances of about $1 \times 10^4 - 5 \times 10^4$ AU. In Fig. 1.7 the outer solar system with these two belts is shown.

Therefore, millions of asteroids exist, many thought to be the scattered remnants of planetesimals, bodies within the young Sun's surrounding nebula that never grew large enough to become planets.

Asteroids are mainly composed of minerals and rock, whereas comets are mainly composed of ice.

A detailed review of the 12 asteroids that have been visited by eight robotic spacecraft is presented in Sears [2015].

The origin and evolution of the main asteroid belt was reviewed by O'Brien and Sykes [2011]. Vesta and Ceres are the largest members of the asteroid belt. The dynamical evolution of the asteroid belt can be

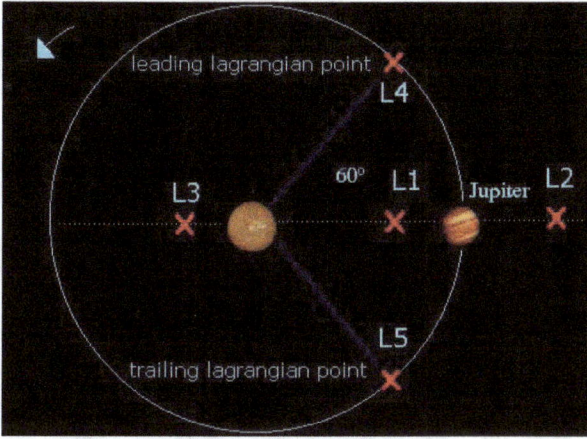

Fig. 1.8 The Lagrange point in the Sun–Jupiter system. L_4 and L_5 form an equilateral triangle with the Sun and Jupiter and are stable points.

reconstructed from counting craters on these objects. A surface with fewer craters must be younger than a heavily cratered surface.

The group of Trojan asteroids are located on the Lagrangian equilibrium points L_4 and L_5 in the Sun–Jupiter system. This is shown in Fig. 1.8.

Furthermore, we can also find ring systems around the four giant planets. Saturn's rings are bright and broad, Jupiter's ring is very tenuous and composed mostly of small particles, Uranus has nine opaque rings and Neptune has four rings. The ring formation around giant planets by tidal disruption of a single passing large Kuiper Belt object is studied in Hyodo *et al.* [2017].

1.1.5 *The heliosphere*

The whole planetary system is enclosed by the heliosphere (Fig. 1.9). The heliosphere is formed by solar plasma (solar wind) and solar magnetic fields. The solar wind particles travel outwards from the Sun at supersonic speeds. At the heliopause, the solar wind merges with the interstellar medium. The density of the solar wind protons and electrons decreases with distance squared. At the Earth's orbit (1 AU), the typical density is about 5 protons per cm^3 and the speed is about 400 km/s. The heliopause is located somewhere between 100 and 200 AU. The location varies with the 11 year solar activity cycle. On September 12, 2013, NASA announced that Voyager 1 had exited the heliosphere on August 25, 2012, when it measured a sudden increase in plasma density of about 40 times. Since the

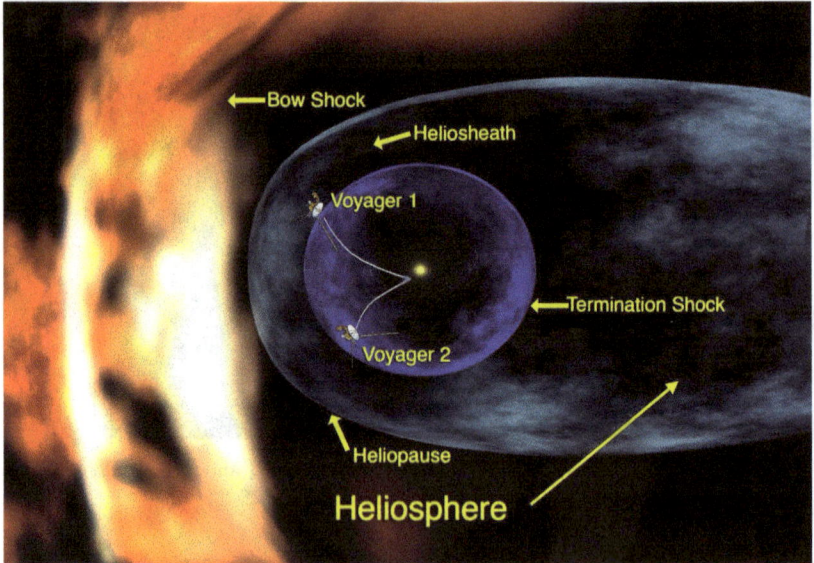

Fig. 1.9   The heliosphere. Credit: NASA.

heliopause marks one boundary between the Sun's solar wind and the rest of the galaxy, a spacecraft such as Voyager 1 that has departed the heliosphere has reached interstellar space (see Borovikov and Pogorelov [2014]).

1.2   Physical parameters of planets

In this section we discuss how the most important physical parameters that characterize planets and other solar system bodies are derived and compare the planets to each other.

1.2.1   *Orbits*

Johannes Kepler (1571–1630) is known for his three laws on planetary motion:

- All planets move along elliptical paths with the Sun at one focus.
- The line connecting any planet and the Sun passes area at a constant rate, i.e. a planet on an elliptical orbit moves faster when it is closer to the Sun.
- If a denotes the semi-major axis of a planetary orbit, T the planet's orbital period, G the gravitational constant, $m_\odot$ the mass of the Sun

and m_p the mass of the planet, Kepler's third law becomes:

$$\frac{a^3}{T^2} = \frac{G}{4\pi^2}(m_\odot + m_p). \tag{1.1}$$

Kepler's third law is extremely useful for the determination of masses. If the orbital period and the semi-major axis of an object are known, the mass of the central body can be calculated. The orbit of a planet is defined by six orbital elements:

- a semi-major axis.
- e eccentricity, for $e = 0$ the orbit is circular, for $e = 1$ parabolic.
- i inclination, tilt of the orbital plane to some reference plane. In the case of the solar system the reference plane is the plane of the Earth's orbit, called ecliptic.
- $\bar{\omega}$ longitude of periapse.
- Ω longitude of the ascending node.
- ν true anomaly.

The orbital elements are shown in Fig. 1.10. The inclination is small for all planets, by definition $i = 0$ for Earth. The planet with the largest

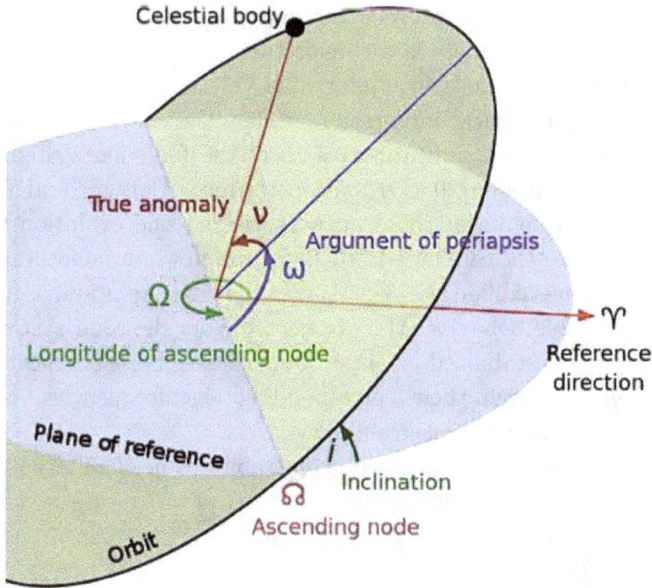

Fig. 1.10   The orbital elements.

inclination is Mercury ($i = 7°$); the dwarf planet Pluto has the inclination $i = 17.1°$. The eccentricity for Earth is 0.016, for Mars $e = 0.09$, for Mercury $e = 0.21$ and for Pluto $e = 0.25$. The eccentricity determines the perihelion r_{min} and aphelion r_{max} distance of a planet: perihelion means that the planet is at its closest position to the Sun.

$$r_{min} = a(1 - e), \tag{1.2}$$

$$r_{max} = a(1 + e). \tag{1.3}$$

Halley's comet has an eccentricity of $e = 0.96$. Being far away from the Sun, solar activity has no influence; however, being close to the Sun the solar activity leads to outgassing of volatile elements. Of the many exoplanets discovered, most exoplanets have a higher orbital eccentricity than our solar system planets. Exoplanets found with low orbital eccentricity are very close to their star and are tidally locked to the star.

Comets, asteroids and minor satellites of planets may have larger inclinations. Some of them move in a retrograde sense (opposite to the Sun's/planet's rotation).

Another important aspect concerns the stability of the planetary system. This can be only verified by using a long-term numerical integration of the equations of motion. The results show that all planets are stable with the exception of Mercury; there is a chance that Mercury will leave the solar system after having a close encounter with Venus.

A long-term numerical integration of the classical Newtonian approximation to the planetary orbital motions of the full solar system (Sun + eight planets), spanning 20 Gyr, was performed [Batygin and Laughlin, 2008]. The experiments of these authors yielded one evolution in which Mercury falls onto the Sun at ∼1.261 Gyr from now, and another in which Mercury and Venus collide in ∼862 Myr. In the latter solution, as a result of Mercury's unstable behavior, Mars is ejected from the solar system at 822 Myr. Mercury is destabilized via an entrance into a linear secular resonance with Jupiter in which their corresponding eigenfrequencies experience extended periods of commensurability.

Thus, resonance effects can lead to instabilities in planetary systems.

1.2.2 *Mass*

The mass of an object can be derived from the law of gravity. Two objects with masses m_1, m_2 attract each other according to Newton's law of gravitation:

$$F = G\frac{m_1 m_2}{r^2}, \qquad (1.4)$$

r being the distance between the two masses, and $G = 6.67 \times 10^{-11}\,\mathrm{m^3\,kg^{-1}\,s^{-2}}$ the gravitational constant. Let us consider some examples for the solar system:

- Planets with moons: the masses of satellites of planets are very small compared to that of their host planet. Therefore, we can apply Kepler's third law.
- The mass can be also deduced from perturbation of orbits. Neptune was discovered since it perturbs the orbit of Uranus. The masses of asteroids can also be found using this method. There are short- and long-term perturbations. Perturbations of the orbits of Neptune and Uranus can be also used to detect possible planets beyond the orbit of Neptune (Planet X). The mean motion of the disturbing body introduces measurable modulations in the residuals. Uranus has been observed over a larger fraction of its orbit; therefore, this planet is nearly two times as sensitive as Neptune to detect an unknown object [Brunini, 1992]. It seems to be that a distance of about 100 AU is the upper limit for such a planet, otherwise the perturbations would become too small.
- Spacecraft tracking. The Doppler shift of the transmitted radio signal can be measured with high precision. Using this method, the mass of the moons of the outer planets was determined.

1.2.3 Size

There are several methods to determine the size of objects in the solar system.

- If the angular diameter is known in radians and the distance in km, the size in km can be derived.
- Occultation of stars: The duration of an occultation at a particular observing site gives the length of a chord of the body's projected silhouette. Three well-separated chords suffice for a spherical planet. For small bodies like asteroids and satellites many chords give their irregular shape.
- Radar echoes: the radar signal drops as $1/r^4$. This drop comes from (i) $1/r^2$ going to object, (ii) $1/r^2$ returning to the antenna. This method works for solid bodies, asteroids and cometary nuclei.

Fig. 1.11 Due to its fast rotation planet Jupiter becomes an ellipsoid. Credit: NASA/Voyager.

- Photometric observations in visible and IR wavelengths. At the visible wavelength one measures the sunlight reflected off the object, at IR wavelength one observes the thermal radiation from the body.

From the diameter and the mass follows the mean density of the objects. Terrestrial planet densities lie between 3.5 and 5.5 g cm^{-3} since they consist of rocky material and metals. The four giant planets have low densities (≈ 1 g cm^{-3}). Comets consist of dirty ice, and their density is also near 1 g cm^{-3}. From the mass M and radius r of an object we can calculate the escape velocity:

$$v_{\mathrm{esc}} = \sqrt{2GM/r}. \tag{1.5}$$

As soon as the thermal velocity of constituents of planetary atmospheres exceeds v_{esc}, these volatiles will escape into space.

1.2.4 *Rotation*

By the observation of surface details or permanent atmospheric features, the rotation of planets can be derived directly (Fig. 1.11).

Planets with magnetic fields trap charged particles within their magnetospheres. These charged particles are accelerated and emit radio waves. The magnetic field rotates with the planet and therefore the radio emission has a periodicity that is equal to the rotation period of the planet.

Lightcurve variations are caused by different albedo (due to clouds in the atmosphere or surface structures) or when the bodies are irregularly shaped

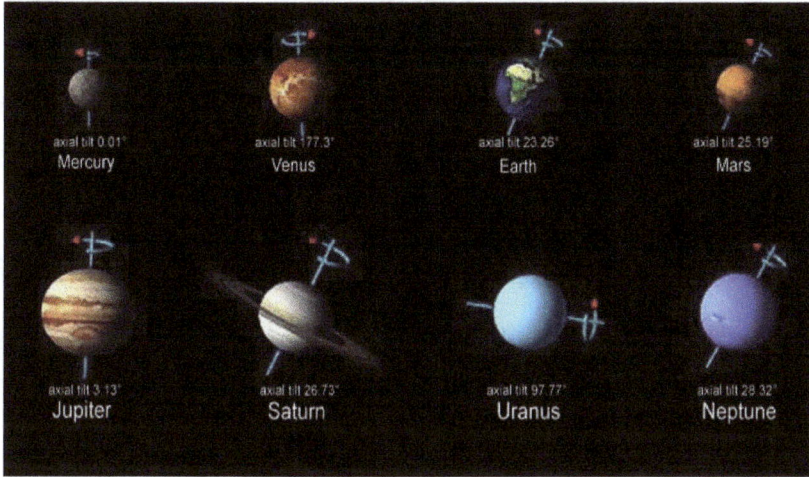

Fig. 1.12 Rotation of the eight solar system planets and axial tilt. Credit: redshiftlive.

(asteroids). Rotation periods can also be derived by measuring Doppler shifts of spectral lines across the disk.

Mercury has a rotation period of 59 days (2/3 of its revolution around the Sun), and Venus has a period of 243 days. Venus rotates in retrograde direction, the rotation axis has an obliquity of 177°. The rotation axis of the eight solar system planets is shown in Fig. 1.12; note the tilt of Uranus and Pluto. A large inclination of the rotation axis can be explained by catastrophic collisions in the young solar system.

Gravity tends to contract a celestial body (larger than 200 km) into a sphere, the shape for which all the mass is as close to the center of gravity as possible. Rapid rotation introduces a centrifugal force, and the planet becomes oblate. Polar flattening is greatest for planets that have a low density and a rapid rotation. This is the case for the giant planets. If a denotes the major axis and b the minor axis of an elliptically shaped planet, the flattening is defined as

$$f = \frac{a - b}{a}. \tag{1.6}$$

For the WGS84[2] ellipsoid to model Earth, the defining values are:

$$a = 6{,}378{,}137.0 \, \text{m},$$
$$b = 6{,}356{,}752.31 \, \text{m}.$$

---

[2]A spatial reference system.

Table 1.2.   Solar system bodies and flattening.

Body	Equatorial bulge (km)
Earth	42.77
Mars	50.2
Ceres	66
Jupiter	10,175
Saturn	11,808
Uranus	1,172
Neptune	846

The difference between the major and minor semi-axes is 21.4 km. The equatorial bulge of the Earth is 42.77 km. In Table 1.2 the equatorial bulge of some solar system bodies is given.

The above defined flattening coefficient can be approximated by

$$f = \frac{5}{4}\frac{\omega^2 a^3}{GM},$$ (1.7)

where a is the mean radius, $\omega = 2\pi/T$ the angular velocity and T is the rotation period. Let us, e.g. consider Jupiter: $a = 6.92 \times 10^7$ m, $\omega = 1.76 \times 10^{-4}$ rad/s, $M = 1.9 \times 10^{27}$ kg. We find $f = 0.101$.

The abovementioned formula is valid under the following assumptions:

- equilibrium configuration of a self-gravitating spheroid;
- planet is composed of an incompressible fluid of uniform density;
- rotation is around a fixed axis;
- for a small amount of flattening.

1.2.5   Temperature

The amount of energy a planet receives every second from the Sun is given by

$$P_{\text{in}} = L_\odot(1 - A)\left(\frac{\pi R_p^2}{4\pi D^2}\right).$$ (1.8)

A denotes the albedo of a planet (fraction of incoming energy that is reflected off the planet). A low value of A denotes a low reflectivity, whereas, for example, ice has a high reflectivity. Jupiter's satellite Europe is covered by an icy crust (albedo 0.68); however, the surface of the Moon is very dark with a low albedo ($A = 0.12$). R_p is the radius of a planet, D denotes its distance from the Sun and $L_\odot$ the luminosity of the Sun.

$L_\odot$ depends on the surface area (given by the radius of the Sun, $R_\odot$) and the Stefan–Boltzmann law (which means it is proportional to T^4):

$$L_\odot = 4\pi R_\odot^2 \sigma T_\odot^4. \tag{1.9}$$

On the other hand, the amount a planet is giving off each second (not reflecting, but emitting) is given by

$$P_{\text{out}} = 4\pi R_p^2 \sigma T_p^4. \tag{1.10}$$

In case of thermal equilibrium, these rates have to be in balance:

$$P_{\text{in}} = P_{\text{out}}. \tag{1.11}$$

And we obtain the equilibrium temperature

$$T_{\text{eq}} = T_\odot (1 - A)^{1/4} \sqrt{\frac{R_\odot}{2D}}. \tag{1.12}$$

The equilibrium temperature of a planet does not depend on the radius of the planet.

However, this is a very crude approximation. Planets may have internal heat sources (such as Jupiter, Saturn or Neptune). If the atmosphere of a planet is more transparent to visible radiation than to infrared radiation (from the planet), the surface temperature of the planet raises above the equilibrium value. This is the so-called greenhouse effect.

- Earth: equilibrium temperature is 255 K, greenhouse effect adds 30 degrees to that value.
- Venus: equilibrium temperature of Venus is 241 K; because of its dense atmosphere, the surface temperature is about 740 K after accounting for the greenhouse effect.

In Table 1.3 the albedo and the incoming solar flux as well as the temperature (equilibrium and surface) are given for some planets and

Table 1.3. Temperatures and Albedo values for some planets and Saturn's satellite Titan.

Planet	Albedo	Solar flux	T_{eq}	T_{surf}	Greenhouse
Mercury	0.1	10134.15	448	448	No
Venus	0.7	2533.54	241	740	Extreme
Earth	0.3	1371.61	255	288	Yes
Mars	0.15	590.82	217	227	Small
Jupiter	0.45	51.70	106	—	—
Titan	0.22	11.66	85	93	Small

Saturn's satellite Titan. Please note that because of the higher albedo, the equilibrium temperature for Venus is lower than for Earth although the Venus–Sun distance is only 2/3 the Earth–Sun distance.

The flux received from the Sun on a planet depends on the total flux on the Sun $F_\odot = \sigma T^4 4\pi R_\odot^2$ and the flux at a sphere of radius of the planet's orbital radius

$$F_{\text{planet}} = \frac{\sigma T_\odot^4 \pi R_\odot^2}{4\pi r_{\text{pl}}^2}. \tag{1.13}$$

The flux is given in W/m^2.

The amount of radiation striking the planet is this flux times the cross-sectional area of the planet

$$F_{\text{tot}} = \sigma T_\odot^4 \frac{R_\odot^2}{r_{\text{pl}}^2} \pi r_{\text{pl}}^2. \tag{1.14}$$

The concept of equilibrium temperature is illustrated in Fig. 1.13.

In Fig. 1.14 the measured temperature profiles in the atmospheres of Venus, Earth and Mars are shown. The temperature profiles shown here can be explained by photodissociation, absorption and ionization processes, e.g. the temperature increase in the Earth's stratosphere by ozone production.

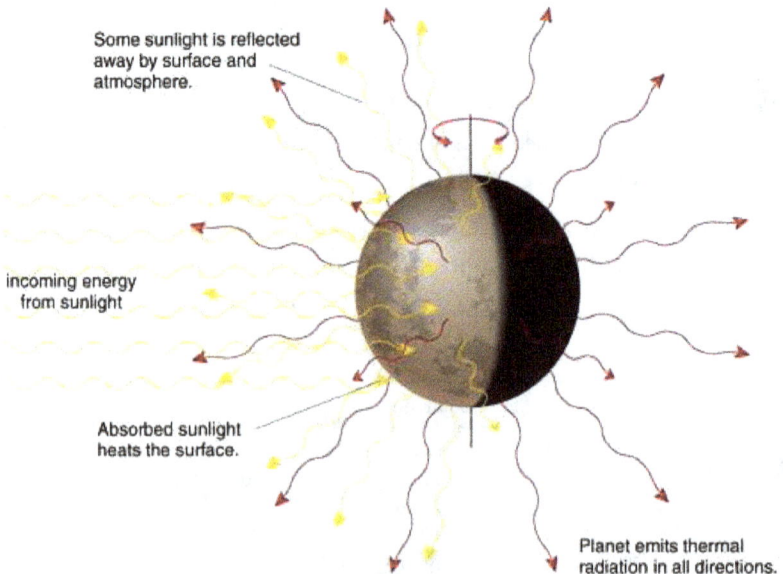

Fig. 1.13   Concept of equilibrium temperature. Credit: lasp.colorado.edu.

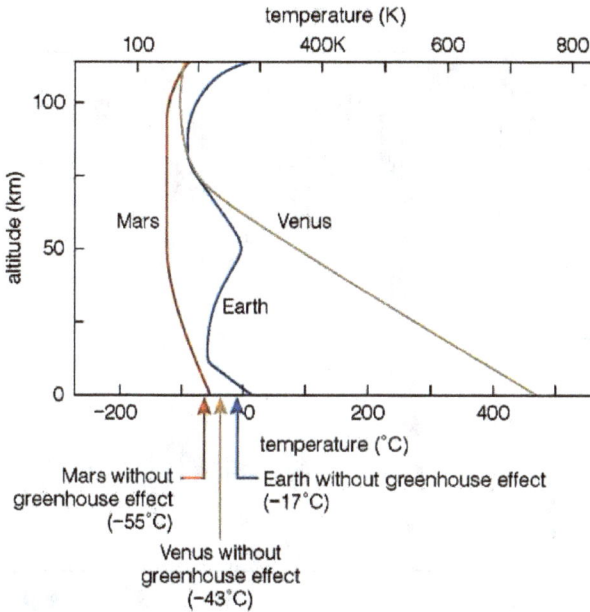

Fig. 1.14 Measured temperature profiles for Venus, Earth and Mars. Credit: lasp.colorado.edu.

The opacity of planetary atmospheres varies with wavelength. This allows us to probe different altitudes in a planetary atmosphere. One can also do photometric measurements in different wavelength bands. In Fig. 1.15 the transparency of the Earth's atmosphere is shown. The letters I, J, H, K, L, M denote the transparent IR windows.

1.3 Planetary magnetic fields

1.3.1 Magnetic fields

Magnetic fields are created by moving charges. Since currents moving in a solid body decay quickly (exception: if the medium is a superconductor), planetary magnetic fields must be produced by (i) a dynamo process or (ii) remnant ferromagnetism. The explanation of planetary magnetic fields by remnant ferromagnetism is not realistic because of the decay on timescales that are shorter than the age of the solar system. Very small magnetic fields may also be induced by interaction between the solar wind (which is composed predominantly of charged particles) and conducting regions on the planet. A dynamo process may only operate in a fluid

Transmission of Earth's atmosphere

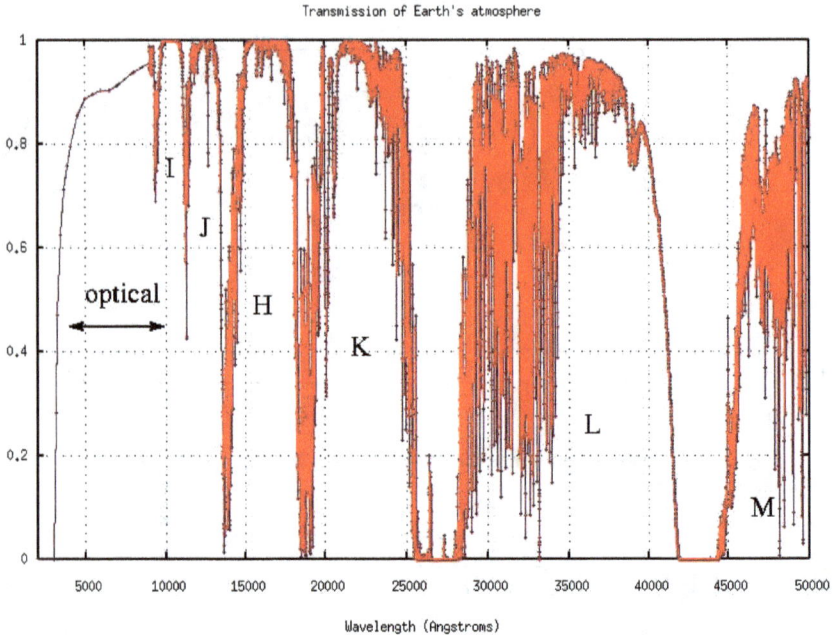

Fig. 1.15   The transparency of the Earth's atmosphere from the optical to the near IR. Credit: M. Richmond.

region of a planet where charges can move in combination with a sufficient planetary rotation rate. This is the case for

- **Terrestrial planets:** liquid Fe–Ni core (e.g. Earth).
- **Gas giants:** a layer of metallic hydrogen exists in the interior, i.e. due to high pressure and temperature, hydrogen has metallic properties i.e. free moving electrons exist.

Planetary magnetic fields can be measured directly by using *in situ* magnetometers. Planetary radio emission can also be a hint for the presence of a magnetic field; the radio emission is due to accelerated charges in the magnetic fields. Localized aurorae also indicate the presence of planetary magnetic fields. An example of aurorae on Jupiter is shown in Fig. 1.16.

The ionospheric response to auroral precipitation on the giant planets and the emission processes for aurorae at radio, infrared, visible, ultraviolet, and X-ray wavelengths are described, and exemplified, using ground- and space-based observations in the review given by Badman *et al.* [2015].

All four giant planets as well as Earth and Mercury have magnetic fields that are generated in their interiors. Mars does not have a global magnetic

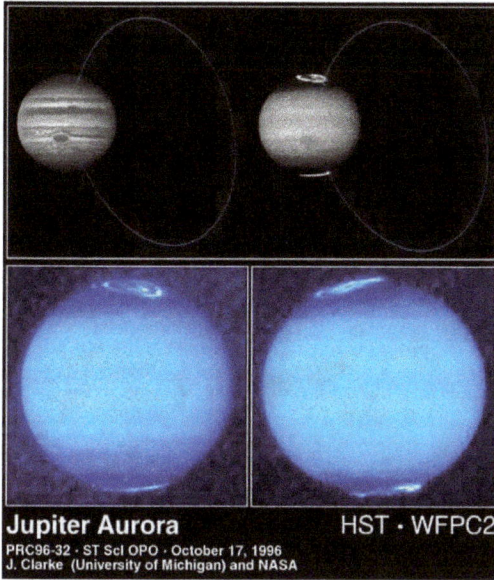

Fig. 1.16 Aurorae on Jupiter. Credit: NASA.

field. Its interior has cooled because of the relatively low mass; Venus has no interior magnetic field because of its low rotation rate. Venus and comets have magnetic fields induced by the interaction between the solar wind and charged particles in their atmospheres/ionospheres.

Induced magnetospheres (IMs) can be found on Venus, Mars and Saturn's moon Titan. All three objects form a well-defined IM and magnetotail as a consequence of the interaction of an external wind of plasma with the ionosphere and the exosphere of these objects [Bertucci *et al.*, 2011]. The IM of Venus is shown in Fig. 1.17.

Mars and the Moon have strong localized crustal magnetic fields.

The interior structure of the planets is shown in Fig. 1.18.

1.3.2 *Geodynamo*

The magnetic field of the Earth can be explained by the geodynamo model. It is generated by the motion of electrically charged particles. The Earth's magnetic field originates in its core, which mainly consists of iron and extends to about 3,400 km. The core is divided into the solid inner core and the liquid outer core. The motion of the liquid outer core is driven by heat flow from the inner core ($T \approx 6{,}000\,\mathrm{K}$). The flow pattern is influenced by the Earth's rotation. The dynamo model of the Earth is shown in Fig. 1.19.

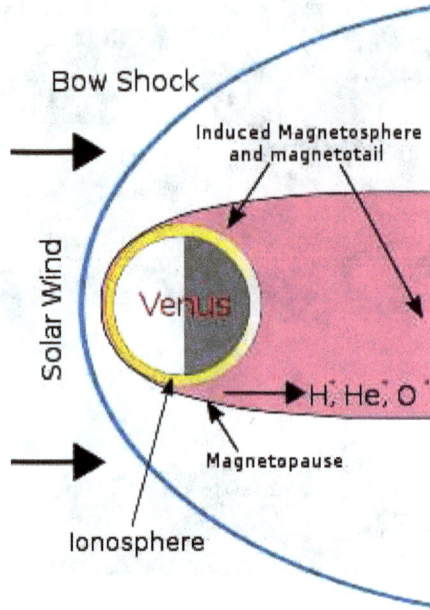

Fig. 1.17   The induced magnetosphere of Venus. Credit: Ruslik.

Close to its surface, the strength of the magnetic field of the Earth can be calculated by assuming it is a simple dipole field, and the strength depends on distance R, geomagnetic latitude λ and on the dipole momentum $M = 7.746 \times 10^{24}\,\text{nTm}^3$:

$$B(R, \lambda) = \frac{M}{R^3}\sqrt{1 + 3\sin^2(\lambda)}. \qquad (1.15)$$

The tilt of the magnetic field of the planets with respect to their rotational axis is quite different from planet to planet, as is illustrated in Fig. 1.20. Please note that in reality the magnetic field strength of e.g. Jupiter is 20,000 larger than that of Earth and it is 100 times more extended.

The induction equation that can be derived from Maxwell's equations is fundamental for the mathematical description of these processes:

$$\frac{\partial \mathbf{B}}{\partial t} = \eta \nabla^2 \mathbf{B} + \nabla \times (\mathbf{v} \times \mathbf{B}). \qquad (1.16)$$

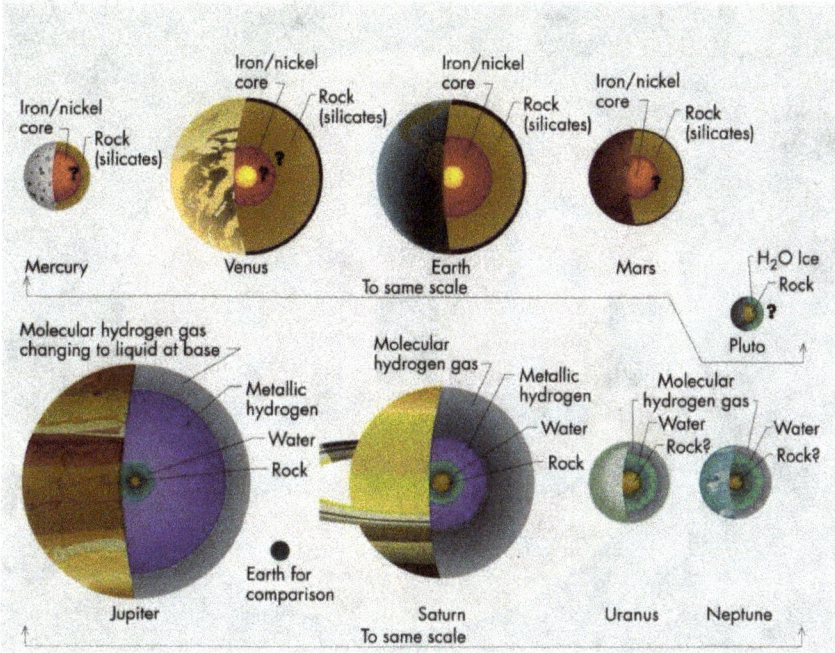

Fig. 1.18 The interior of planets. Credit: NASA.

$\eta = 1/\mu_0\sigma$ is the magnetic diffusivity, and σ is the electrical conductivity. Let us assume the fluid moves with a typical speed V and a typical length scale L.

$$\eta\nabla^2\mathbf{B} \sim \frac{\eta B}{L^2}, \tag{1.17}$$

$$\nabla \times (\mathbf{v} \times \mathbf{B}) \sim \frac{VB}{L}. \tag{1.18}$$

From the ratio of these quantities, the magnetic Reynolds number can be derived:

$$R_m = \frac{LV}{\eta}. \tag{1.19}$$

In a typical star there is infinite electrical conductivity, $\eta \to 0$; thus, a huge Reynolds number results and the induction equation becomes

$$\frac{\partial\mathbf{B}}{\partial t} = \nabla \times (\mathbf{v} \times \mathbf{B}). \tag{1.20}$$

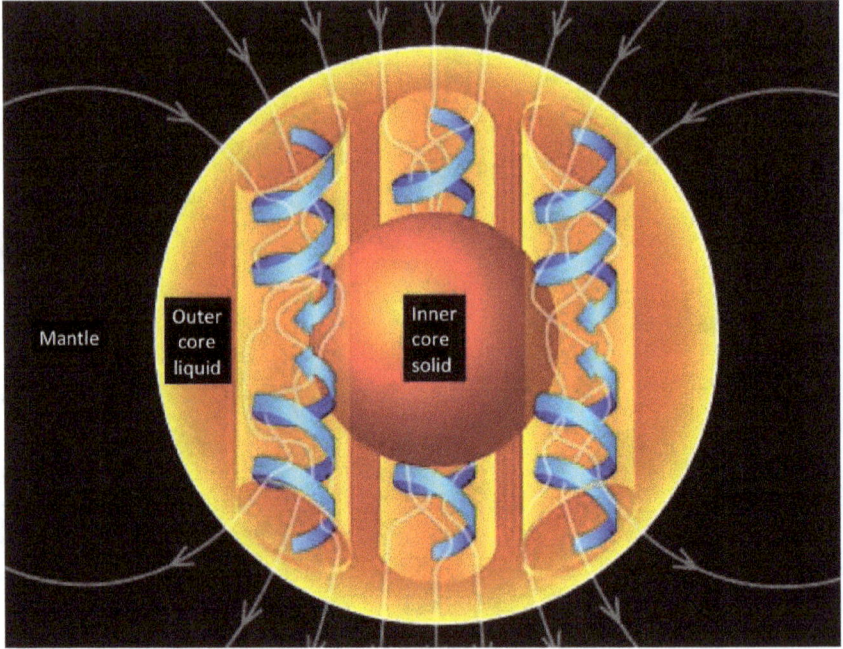

Fig. 1.19 The geodynamo. Credit: NASA.

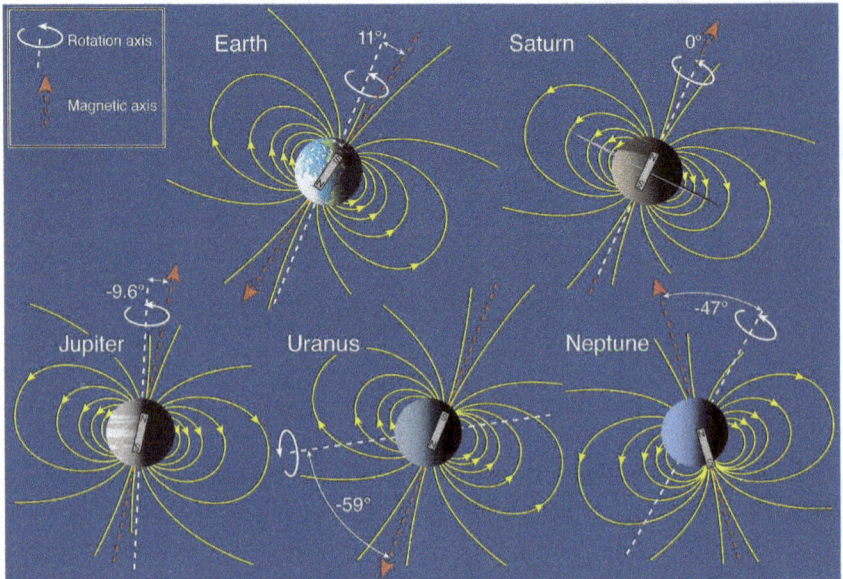

Fig. 1.20 Tilt of the magnetic field of planets, not in scale. Credit: NASA.

For small Reynolds numbers, the diffusive term overcomes the convective term and

$$\frac{\partial \mathbf{B}}{\partial t} = \eta \nabla^2 \mathbf{B}. \tag{1.21}$$

Summary: When does a planet have a magnetic field?

- For terrestrial planets:

  — liquid interior,
  — sufficient rotation rate.

- For gas planets:

  — high pressure, metallic hydrogen, free electrons moving,
  — rapid rotation.

Observations and models of planetary magnetic fields are reviewed in Schubert and Soderlund [2011]. Meteorites and asteroids can inform us about magnetic fields in the epoch of their formation. Most small moons and small planets have a residual, random, or remanent magnetism [Vallée, 2011]. The magnetic field of the Earth is by far the best documented magnetic field of all known planets (see review by Hulot *et al.* [2010]).

1.3.3 *Shielding effects*

The existence of a planetary magnetic field is essential because charged solar wind particles are deflected and cannot penetrate the field lines. Therefore, the planetary surface is protected from a bombardment of highly charged energetic particles and life on the surface of Earth is possible. The second shielding factor against solar influences comes from the atmosphere.

The Earth's magnetosphere becomes strongly deformed by the influence of solar wind. On the side facing the Sun it gets strongly compressed, at the opposite side we find the so-called magnetotail (see Fig. 1.21).

Jupiter produces a plasma torus from its satellite Io, which is volcanically very active and expels particles that are trapped in the plasma torus by Jupiter's magnetic field.

The strengths of the magnetic fields relative to the strength of the Earth's field is given in Table 1.4. The total strength (NSSDC gives strength in gauss $\times R_p^3$, where R_p is the radius of the planet and Earth's strength $= 0.3076$ G $\times R_\oplus^3 = 7.981 \times 10^{10}$ G).

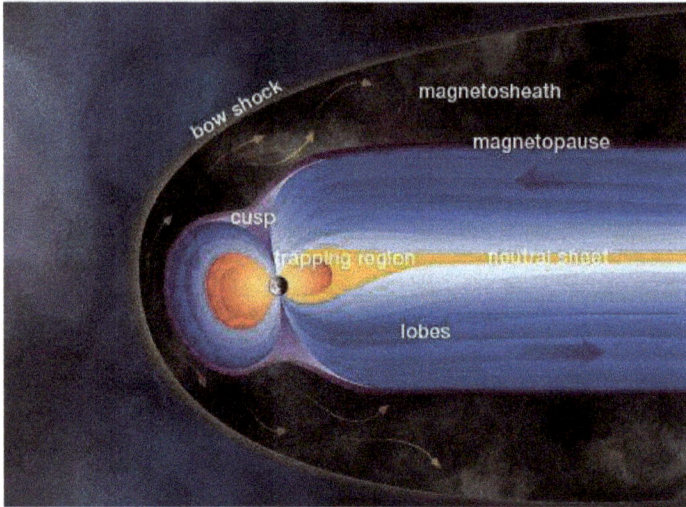

Fig. 1.21   The Earth's magnetosphere. Credit: NASA.

Table 1.4.   Relative strengths of the magnetic fields of the solar system planets.

Planet	Magn. field (given in G $\times R_p^3$, R_P planet's radius)
Mercury	0.006
Venus	0.00
Earth	1
Mars	0.00
Jupiter	19,519
Saturn	578
Uranus	47.9
Neptune	28.0

1.4   The atmospheres of planets

1.4.1 *Basic parameters describing an atmosphere*

An important parameter for a planetary atmosphere is the scale height. This is the height at which the atmospheric pressure declines by a factor of $e = 2.71828$. The scale height is often denoted by H. For an atmosphere with uniform temperature, the scale height is proportional to T and inversely proportional to the product of the mean molecular mass and the local acceleration of gravity; the pressure declines exponentially with

Table 1.5. Scale height for planets and Saturn's satellite Titan.

Planet	Scale height (km)
Venus	15.9
Earth	8.5
Mars	11.1
Jupiter	27
Saturn	59.5
Titan	40
Uranus	27.7
Neptune	≈ 20

increasing altitude.

$$H = \frac{kT}{Mg} \tag{1.22}$$

with $k = 1.38 \times 10^{-23}\,\mathrm{J\,K^{-1}}$.

Since the pressure gradient can be written as

$$\frac{dP}{dz} = -g\rho, \tag{1.23}$$

we obtain

$$\frac{dP}{P} = -\frac{dz}{H}, \tag{1.24}$$

$$P = P_0 \exp\left(-\frac{z}{H}\right). \tag{1.25}$$

In Table 1.5 we give the approximate scale heights for planets and the Saturn satellite Titan.

Surface gravity g is an important parameter for a planet in order to have an atmosphere.

$$g = \frac{GM}{R^2}. \tag{1.26}$$

For different solar system bodies, the surface gravity relative to the Earth's is shown in Table 1.6. The Earth's surface gravity is $g = 9.81\,\mathrm{m/s^2}$.

In this table, Deimos and Phobos are the two tiny satellites of Mars (diameter only several tens of km), Triton is Neptune's largest satellite.

Another important factor is the distance of a planet from the Sun. The smaller the distance, the larger are the thermal motions of the gas due to higher solar insolation. The average kinetic energy of a particle is

$$\frac{1}{2}m\bar{v}^2 = \frac{3}{2}kT. \tag{1.27}$$

Table 1.6.   Surface gravity of some solar system objects.

Object	Surface gravity (g)	Object	Surface gravity (g)
Sun	28.02	Io	0.183
Mercury	0.38	Europa	0.134
Venus	0.904	Ganymede	0.15
Earth	1.00	Callisto	0.126
Moon	0.165	Titan	0.14
Mars	0.376	Enceladus	0.0113
Jupiter	2.53	Triton	0.0797
Saturn	1.07	Pluto	0.067
Uranus	0.89	Phobos	0.0005
Neptune	1.4	Deimos	0.0003

The root mean square (rms) speed is

$$v_{\mathrm{rms}} = \sqrt{\frac{3kT}{m}}. \tag{1.28}$$

The most probable speed is 0.82 of the rms speed.

1.4.2   *Atmospheric escape*

If the thermal motion of the gas particles exceeds the escape speed, the particles escape from the atmosphere. The escape speed is defined by

$$v_{\mathrm{esc}} = \sqrt{\frac{2GM}{R}}, \tag{1.29}$$

where M is the mass and R the radius of a planet. The velocities of the particles are distributed according to a Maxwell–Boltzmann distribution $f(v)$

$$f(v) = \left(\frac{m}{2\pi kT}\right)^{3/2} 4\pi v^2 \exp\left(-\frac{mv^2}{2kT}\right). \tag{1.30}$$

Molecules in the tail of the distribution may reach the escape velocity. If this occurs at an atmospheric height where the mean free path of a particle is equal to the scale height, the particles will escape. This process is also known as *Jeans escape*. The large gravitational force of the giant planet Jupiter is able to retain light gases such as hydrogen and helium that escape from objects with lower gravity.

Lighter molecules move faster than heavier ones with the same thermal kinetic energy; hence, gases of low molecular weight are lost more rapidly than those of high molecular weight. It is thought that Venus and Mars may have lost much of their water when the atmospheric hydrogen escaped after H_2O photodissociated into hydrogen and oxygen by solar ultraviolet radiation; the hydrogen escaped. The Earth's magnetic field helps to prevent this, as, normally, the solar wind would greatly enhance the escape of hydrogen. However, over the past 3 billion years Earth may have lost gases through the magnetic polar regions due to auroral activity, including a net 2% of its atmospheric oxygen.

In Fig. 1.22 the planets' escape velocity, their surface temperatures, and the gases and molecules that can be held in their atmospheres are shown.

Another process is *hydrodynamic escape*. Here, the atmosphere flows off like particles in a comet's tail. Pressure gradients induced by thermal energy deposition lead to this dramatic loss of an atmosphere. Through drag effects the lighter escaping particles also drive off heavier atoms. Exoplanets

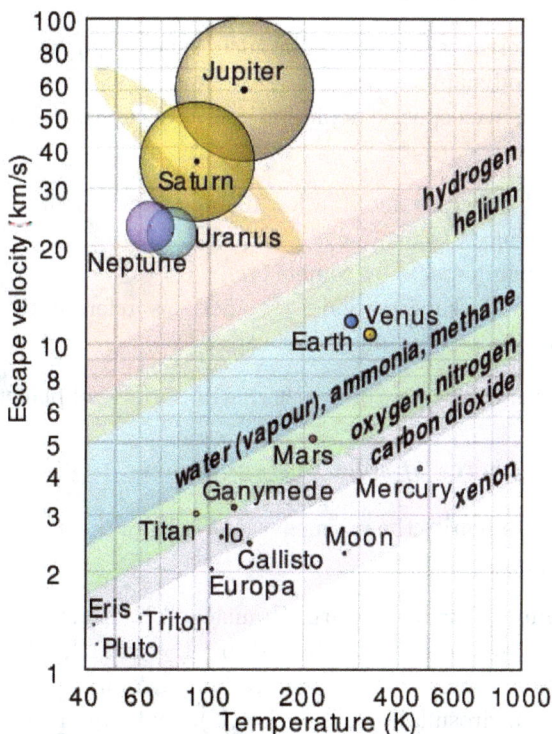

Fig. 1.22 Graphs of escape velocity against surface temperature of some solar system objects showing which gases are retained. The objects are drawn to scale, and their data points are at the black dots in the middle. Data is taken from http://ircamera. as.arizona.edu/astr_250/Lectures/Lec_05sml.htm and http://www-spof.gsfc.nasa.gov/ stargaze/StarFAQ12.htm.

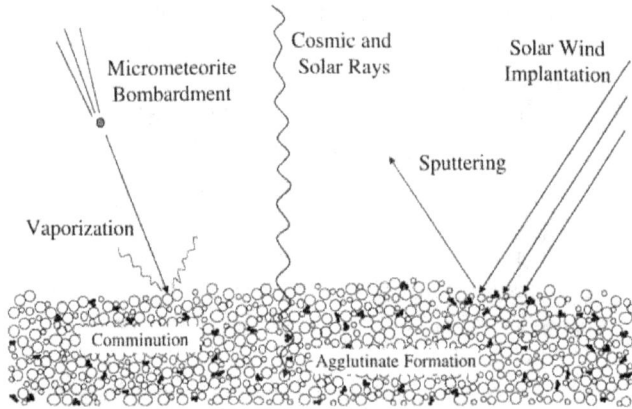

Fig. 1.23 Cartoon showing the different components of space weathering. Credit: Wikimedia commons, IntrplnetSarah.

that are extremely close to their star, such as hot Jupiters, can experience significant hydrodynamic escape.

In summary, atmospheric depletion can have several causes:

- low surface gravity;
- high thermal velocities due to a low distance from the host star;
- solar wind-induced sputtering;
- erosion through impacts by comets etc.;
- weathering, and sequestration — sometimes referred to as "freezing out" — into the regolith and polar caps.

In Fig. 1.23 several processes of weathering on surfaces of planets are shown.

1.4.3 *Atmospheres of solar system planets*

Let us briefly summarize the main characteristics of the atmospheres of the solar system planets.

- **Giant planets:** Jupiter, Saturn, Uranus and Neptune. Huge fluid balls, atmospheres are mainly dominated by H_2 and He. In the case of Jupiter, the following cloud layers are found: Ammonia clouds (150 K), Ammonium hydrosulfide clouds (200 K), and Water clouds (270 K). For example, the ammonia clouds on both Jupiter and Saturn form at atmospheric temperature of about 150 K. This happens to be around 25 km below the cloud top on Jupiter and 100 km below the cloud top on Saturn. Methane clouds are formed at a temperature of 75 K.

Methane can condense in the very cold upper troposphere of Uranus and of Neptune, but not in the warmer troposphere of Jupiter or Saturn.

- **Venus:** Very dense CO_2 dominated atmosphere. We cannot see its surface at visible wavelengths.
- **Earth:** N_2 (78%) and O_2 (21%). Without replenishment by photosynthesis, the free oxygen in the Earth's atmosphere would disappear. The lowest atmospheric layer in which convection and weather occurs is the troposphere; in the stratosphere, temperature increases with altitude as a result of the Sun's ultraviolet radiation absorption by the ozone. The ionosphere is a region of Earth's upper atmosphere, from about 60 to 1,000 km altitude, and includes the thermosphere and parts of the mesosphere and exosphere. It is ionized by solar radiation, and plays an important part in atmospheric electricity forming the inner edge of the magnetosphere. Solar shortwave radiation and charged particles mostly affects the ionosphere, and to a small degree the stratosphere.
- **Mars:** Very tenuous CO_2 atmosphere.
- **Titan:** Saturn's largest satellite has a dense nitrogen-rich atmosphere and also organic molecules are found.
- **Pluto and Triton** (Neptune's largest satellite): Have a tenuous N-rich atmosphere.
- **Io:** This satellite of Jupiter is volcanically very active and its tenuous atmosphere consists primarily of SO_2 gas.
- **Mercury:** Very thin atmosphere ($<10^{-12}$ bar), dominated by O, Na and He.
- **Moon:** Very thin atmosphere ($<10^{-12}$ bar), dominated by He and Ar.

1.4.4 *Space weather*

The influence of solar radiation and charged particles on the magnetosphere and atmosphere of a planet is denoted as *space weather*. In Plainaki *et al.* [2016], the scientific aspects of planetary space weather in different regions of our solar system are given; the authors performed a comparative planetology analysis that includes a direct reference to the circum-terrestrial case.

What are the main factors influencing on space weather around planets?

- The distance of the body from the Sun, determining the properties of solar wind and SEPs, the solar photon flux and the properties of the Interplanetary Magnetic Field (IMF) at that location;
- the presence (or absence) of a dense atmosphere;

- the optical thickness (wavelength dependent) of the body's atmosphere and the mean free paths for neutrals and ions traveling therein;
- in case of the giant planet moons, the existence of a strong (or weak) magnetosphere in which the body is embedded (e.g. Europa and Ganymede; Titan and Enceladus versus Miranda and Ariel; Naiade, Talassa and Triton);
- existence of an intrinsic or induced magnetic field;
- the existence of endogenic sources (e.g. plumes, volcanoes);
- galactic cosmic rays (GCRs);
- micrometeorites and dust.

Therefore, space weather acts very differently on the planets. Mercury is strongly exposed both to radiation and particles from the Sun. This planet can be used as a proxy for exoplanets that are close to their parent star. Venus has a very dense atmosphere but no intrinsic magnetic field. Mars has a tenuous atmosphere and no global magnetic field. Space weather on the Galilean Satellites is dominated by the influence of the Sun and Jupiter.

1.4.5 *Evolution of planetary atmospheres*

The whole planetary system formed out of a protoplanetary disk. How did the planets obtain their initial atmospheres? In the solar system, there are four giant planets with massive H- and He-envelopes. For the two gas giants, Jupiter and Saturn these envelopes contain about 85% respectively 60% of the whole planet's mass. For the two ice giants, Uranus and Neptune, the mass of the H- and He-envelopes is about 10% of their total mass. The terrestrial planets Earth, Venus, Mars and Mercury retain no such atmospheres.

Growing protoplanets inside a a protoplanetary disk can be characterized by two parameters: The Hill radius, R_H and the Bondi radius, R_B. Outside the Hill radius, the stellar tidal force is greater than the protoplanet's gravity; outside the Bondi radius, the thermal motion of the gas molecules dominates the protoplanet's gravity. M_p and M_* are the masses of the protoplanet and its host star, a is the planet's orbital distance, μ, γ and T are the mean molecular weight, the ratio of specific heats and the temperature of the protostellar gas cloud, m_u is the atomic mass unit. The formulae for the Hill and Bondi radii are

$$R_H = \left(\frac{M_p}{3M_*} \right)^{1/3} a, \qquad (1.31)$$

$$R_B = \frac{\gamma - 1}{\gamma} \frac{GM_p \mu m_u}{kT}. \qquad (1.32)$$

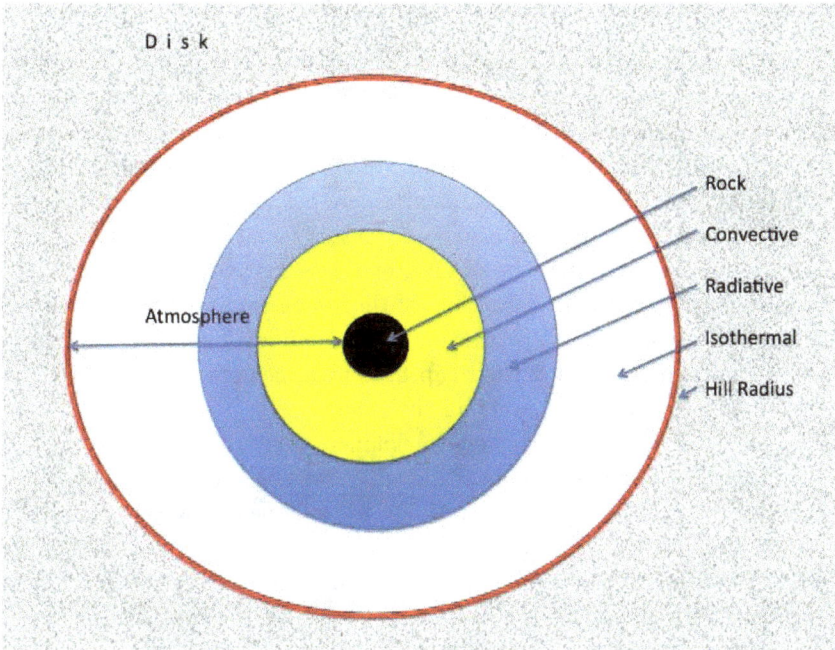

Fig. 1.24 Protoplanetary atmosphere.

An atmosphere can be gravitationally bound (hydrostatic stage) if located (i) inside the Hill radius and (ii) inside the Bondi radius. A formula for the mass M_p of the protoplanet that is required for the existence of a hydrogen atmosphere can be given by

$$M_p \geq 1.1 \times 10^{23} \left(\frac{T}{300\,\mathrm{K}} \right)^{3/2} \left(\frac{\mu}{2} \right)^{-3/2} \left(\frac{\bar{\rho}}{5.5\,\mathrm{gcm}^{-3}} \right)^{-1/2} \mathrm{kg}. \quad (1.33)$$

A schematic view of a protoplanetary atmosphere is shown in Fig. 1.24.

The atmospheres of the giant planets are called primary atmospheres because they contain 10% of He, which is similar to the composition of the protoplanetary disk. The atmospheres of the terrestrial planets are called secondary atmospheres since they have changed during the planet's evolution. The atmospheres of Venus and Mars contain 95% CO_2 and 5% N_2. That would be the same for the Earth; however, (i) CO_2 is now stored in carbonate rocks, and (ii) O_2 is produced by vegetal life.

The following scenarios can be established:

- A protoplanetary core accretes mass and grows. Its gravitational potential increases and it can collect gases like H_2 and other main constituents of the protoplanetary disk. As the processes continues, the protoplanetary disk vanishes leaving a planet with a H_2, He-envelope. At the early stage of stellar evolution, the stars are extremely bright in the UV and X-ray. The envelope of the planets is exposed to this radiation, and atmospheric escape starts. Lower mass bodies accreted less gas and, depending on the intensity of the UV and X-ray flux of the young star, they could lose the gaseous envelope after tens or hundreds of million years. Massive cores however did not get rid of their captured hydrogen envelopes and remain as sub-Neptunes or gas giants all the time.
- Secondary atmosphere: terrestrial planets may have lost their gas envelopes by thermal escape and due to volcanic eruptions; thus, thermally outgassed steam atmospheres developed. These atmospheres consist mainly of CO_2. Planets within the so-called habitable zone solidify within several million years. The outgassed steam atmospheres cool fast, condensation of water and the formation of a liquid ocean sets in.
- In other cases, the magma oceans could remain 100 million years or longer, hydrodynamic atmospheric escape leads to a desiccation of theses planets because the solidification processes were too slow.
- In addition to the abovementioned process of outgassing, the bombardment by meteorites could also have contributed to the evolution of atmospheres.

These processes are described in Massol *et al.* [2016]. To summarize, the followings steps play a crucial role in the formation of planetary atmospheres:

(1) nebular origin of protoatmospheres;
(2) magma ocean outgassed atmospheres;
(3) escape from protoatmospheres.

The formation of planets and the solar system is reviewed in Armitage [2007].

1.4.6 *Venus and Mars: Influence from long-term solar evolution*

Venus is similar to Earth in mass and density. Because of its closer distance to the Sun, the incident sunlight on Venus is 1.9 times that on Earth. Due to the high albedo of the sulfuric acid clouds and massive CO_2 atmosphere (the albedo is about 9%), Venus absorbs less solar flux power per unit area. When comparing Venus to Earth, profound differences can be found. There are no oceans on the surface of Venus and only a very small amount of water vapor in the Venusian atmosphere has been detected. Its surface temperature is 730 K and is much warmer than the critical temperature of water. In Table 1.7 the mass of volatiles per mass of planet are given. The units are mass of volatile per unit mass of planet. The total amount includes the atmosphere, the oceans, frost and minerals (e.g. carbonates in the crust and mantle).

On Earth, carbonates have formed from CO_2 gas dissolving in the ocean. Weathering of he calcium- and magnesium-bearing igneous rocks to form calcium and magnesium carbonates also played an important role. Therefore, the Earth's atmosphere was purified from these compounds. On Venus, this did not happen. The total CO_2 on Earth is comparable to the CO_2 in the Venusian atmosphere (where the pressure is 90 bars). An important proxy for planetary outgassing is ^{40}Ar. This is formed by radioactive decay of ^{40}K in crustal rocks. As it can be seen from Table 1.7, the amount of ^4Ar is comparable for all three planets.

The table also clearly illustrates the big difference in water content between these planets. The explanation as to why Venus has lost most of its water inventory is a runaway greenhouse effect. Water vapor is a potent greenhouse gas. In the Earth's atmosphere there is an equilibrium with the oceans. A warmer ocean means more water vapor and a stronger greenhouse effect, so the temperature increases. Therefore, a positive feedback exists. Water vapor in a planetary atmosphere is vulnerable to photodissociation. UV radiation from the Sun splits the water molecules

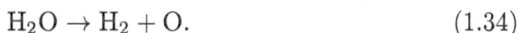

$$H_2O \rightarrow H_2 + O. \tag{1.34}$$

Table 1.7. Comparison of atmospheric composition of Venus, Earth and Mars.

Content	Venus	Earth	Mars
H_2O_{atm}	1.2×10^{-9}	$\leq 1.6 \times 10^{-8}$	$\leq 1.6 \times 10^{-12}$
H_2O_{tot}	1.2×10^{-9}	2.3×10^{-4}	$(3.5 - 5) \times 10^{-6}$
$CO_{2,atm}$	9.5×10^{-5}	5.0×10^{-10}	3.7×10^{-8}
$CO_{2,tot}$	9.5×10^{-5}	$(5 - 15) \times 10^{-5}$	$(6 - 7) \times 10^{-8}$
$N_{2,atm}$	22×10^{-7}	6.7×10^{-7}	6.7×10^{-9}
$O_{2,atm}$	7×10^{-10}	21×10^{-7}	9.5×3.7^{-11}
$^{40}Ar_{atm}$	6×10^{-9}	11×10^{-9}	0.57×10^{-9}

The light hydrogen molecules escape into space and the remaining O and OH combine with the iron in the surface rocks. This runaway greenhouse effect has occurred on Venus, because of its closer distance to the Sun, but not on Earth. The Earth is too far away from the Sun for such an instability to develop.

Further evidence for such a scenario comes from the investigation of the D/H ratio (deuterium to hydrogen ratio). The D/H ratio of Mars is seven times that of Earth, for Venus it is 150 times larger. This can be easily explained. The lighter hydrogen escapes faster than D, leaving more D behind. The explanation of the different evolution of Venus and Earth and the loss of the water on Venus was reviewed by Kulikov *et al.* [2006].

As already discussed, there is evidence for a warm wet early Mars. From stellar evolution theory and the observation of Sun-like stars it follows that the solar luminosity during the first billion years was 20–30% lower than today's value. This is known as the faint young Sun paradox. How could there have been a warmer and wet early Mars when the Sun was fainter then? We will address this point in the next chapters.

Chapter 2

The Sun: The Star We Live With

The Sun is the only star on which details can be observed directly. Therefore, it is an important proxy for stellar astrophysics. Moreover, we can study in detail how solar energetic phenomena like flares, solar wind or CMEs affect on the planets.

2.1 Basic solar parameters and energy output

The value for typical parameters of stars, such as their radius, mass and luminosity, are often in units of the solar value which is denoted as $\odot$.

2.1.1 *Solar radius and mass*

As seen from Earth, the apparent diameter of the Sun in the sky is about 0.5°. The absolute value for the solar radius is:

$$R_\odot = 696{,}000 \, \text{km}. \tag{2.1}$$

The measured solar equator-to-pole radius difference converges toward $8 \, \text{mas}$[1] (corresponding to $5.8 \, \text{km}$). An attempt has also been made to use space-based observations for a more accurate determination of these values (Meftah *et al.* [2015]). The tidal effect of the planets is weak and does not significantly affect the shape of the Sun.

The mass of a star is the basic parameter that determines its evolution, lifetime, stellar activity and other properties. Stellar masses also determine the location of the habitable zone, the so-called circumstellar habitable

---

[1]1 mas = 1 milliarcsec = 0.001 arcsec.

zone. The mass of the Sun can be easily derived from Kepler's third law:

$$M_\odot = 2 \times 10^{30} \, \text{kg}. \tag{2.2}$$

This corresponds to 330,000 times the mass of Earth.

The Sun is constantly losing mass. Solar wind, coronal mass ejections (CMEs, eruptive phenomena expelling plasma to the interplanetary space) as well as the conversion of mass to energy by central nuclear fusion of hydrogen into helium contribute to a loss in mass. Since its formation about 4.5 billion years ago, the Sun may have lost in total about $10^{-3} \, M_\odot$, which is more or less negligible. From the observation of high mass loss rates in very young solar-like stars, we conclude that the solar mass loss was considerably higher during the young phase of solar evolution. Angular momentum and mass loss from magnetized solar-like winds in cool stars are studied in the paper by Pinsonneault *et al.* [2013]. The angular momentum and mass loss of stars during their evolution strongly affects the activity of the stars and also on the evolution of planetary orbits around them.

In Iorio [2010] the influence of solar evolution on planetary orbits that will occur during the next four billion years is estimated. We will discuss these aspects in the chapter about habitability and stellar evolution.

2.1.2 The temperature of the Sun

Compared to other stars, the Sun belongs to the group of cool stars. Its surface effective temperature T_{eff} is about 5,800 K. The effective temperature is obtained by comparing the solar spectral energy distribution with a Planck curve of a black body

$$E(\lambda, T) = \frac{2hc^2}{\lambda^5} \left(\exp(hc/k\lambda T) - 1 \right)^{-1}. \tag{2.3}$$

$h = 6.626 \times 10^{-34} \, \text{J s}$ is the Planck constant.

If we integrate the Planck function over all wavelengths, we obtain the Stefan–Boltzmann law, which defines the effective temperature T_{eff}:

$$E = \sigma T_{\text{eff}}^4. \tag{2.4}$$

2.1.3 The solar spectrum

In Fig. 2.1 the solar spectrum is shown. As it is seen in this figure, different parts of the solar spectrum are absorbed in the Earth's atmosphere. Short-wavelength radiation (e.g. UV) is absorbed by ozone, O_3, in the ozone layer that is found above the troposphere. Therefore, an energetic outburst, e.g. a flare that produces a strong increase in the UV and X-ray part of the solar spectrum, influences these layers.

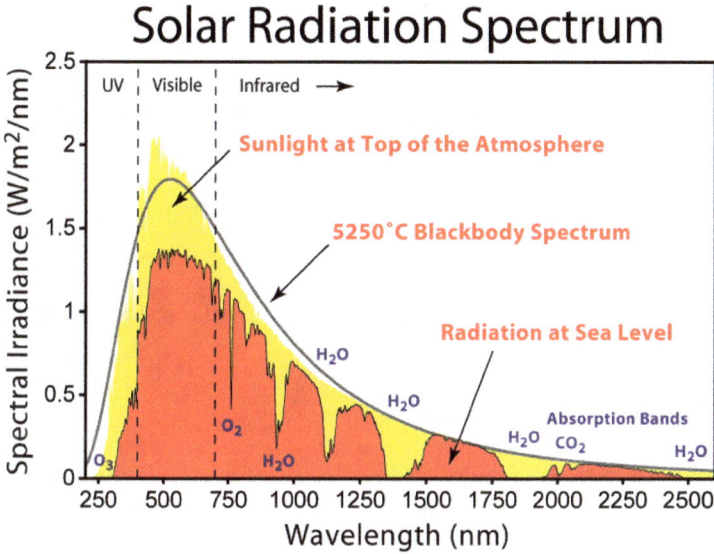

Fig. 2.1 The solar spectrum in the visible range. Credit: NASA.

2.1.4 *Solar luminosity, solar constant*

The solar luminosity is the overall energy output, which means we have to sum up its contribution to radiation at all wavelengths. This energy output is the main driver of climate on the planets. The luminosity of a star generally depends on its total radiating surface and on the integrated Planck function

$$L = 4\pi R^2 \sigma T_{\text{eff}}^4. \qquad (2.5)$$

R denotes the radius of the Sun and σ is the Stefan–Boltzmann constant. When inserting the correct values of the Sun, we obtain the value for the solar luminosity,

$$L_\odot = 3.828 \times 10^{26}\,\text{W}. \qquad (2.6)$$

The total solar irradiance at Earth's distance from the Sun, TSI, has an average value of

$$S_\odot = 1361\,\text{W/m}^2. \qquad (2.7)$$

This value varies slightly because of the

- elliptical orbit of Earth;
- the 11-year solar cycle (0.04%);

- 0.1% fluctuations on timescales of centuries;
- long-term increase of solar luminosity because of solar evolution. In about 4 billion years, the Sun will evolve into a red giant; its radius will increase by a factor of 100 and therefore the luminosity will increase by a factor of 10,000 (assuming that the temperature remains constant). The early Sun was probably fainter than now.

2.2 From the solar interior to the interplanetary space

In this section we discuss the different layers of the solar interior and its atmosphere.

2.2.1 *Hydrostatic equilibrium*

Stars like the Sun are stable because the hydrostatic equilibrium is valid for all layers in the interior. The derivation of the hydrostatic equilibrium equation can be found in all astrophysics textbooks, and so we give only the equation here:

$$\frac{dP}{dh} = -g\rho. \tag{2.8}$$

The gravitational acceleration g is

$$g = \frac{GM}{r^2}. \tag{2.9}$$

The equation for an ideal gas is a good approximation for sun-like stars

$$P = NkT. \tag{2.10}$$

These equations yield a central temperature of about 15 million K.

2.2.2 *Nuclear fusion*

There are three main processes involved:

$$p + p \rightarrow D + e^+ + \nu_e, \tag{2.11}$$

$$p + D \rightarrow {}^3\text{He}, \tag{2.12}$$

$${}^3\text{He} + {}^3\text{He} \rightarrow {}^4\text{He} + 2p. \tag{2.13}$$

In the first reaction, two protons, p, come close to each other to fuse into deuterium, D, which consists of one proton and one neutron. Therefore, one proton decays into (inverse β decay) a positron e^+ and an electron neutrino ν_e.

In the second reaction, a proton reacts with a deuterium and forms the ^{3}He isotope which consists of two protons and one neutron. In the third reaction, two ^{3}He isotopes fuse into one ^{4}He, and two protons are left.

Comparing the mass of four individual protons with the end product ^{4}He, there is a mass difference Δm which is about 0.7%. According to Einstein's formula

$$E = mc^2, \tag{2.14}$$

this Δm is converted into energy E. We can easily show, that by nuclear fusion, the Sun produces enough energy to survive more than 9 billion years as a so-called main sequence star.

The neutrinos that are produced during the fusion processes listed above and other processes not listed can be used to test the models. Neutrinos have a small interaction with the rest of matter and therefore they can penetrate the rest of the Sun and even pass through the Earth without any considerable interaction. Neutrinos are electrically neutral particles; they are not affected by the electromagnetic force and they have a very small rest mass and oscillate between three states: electron neutrinos ν_e, muon neutrinos ν_μ and tau neutrinos, ν_τ.

About 6.5×10^{10} neutrinos from the Sun pass through every square centimeter perpendicular to the direction of the Sun on Earth.

2.2.3 *Radiative and convection zone*

Energy is produced in the solar core by thermonuclear reactions in the form of high energetic gamma ray photons. This energy in the form of short-wavelength photons must be transported to the solar surface. In the radiative zone, the energy is transported by radiation. A high energetic gamma ray photon collides with another particle and is re-radiated again. By these processes, the energy is transported outward and the short-wavelength photons are converted into longer wavelengths. However, it takes about 100,000 years for a gamma ray photon to get transported outward to the solar surface. As a consequence of this, the radiation from the Sun we observe today was produced about 100,000 years ago.

The radiative zone ranges from 0.3 to 0.6 solar radii. The extension of the convective zone ranges from about 0.6 solar radii to the surface; energy is transported by convection there. On the solar surface, we observe a cellular-like structure that is called granulation. In the bright granules, that have an extension of about 1,000 km, hot plasma moves upward, cools and sinks down into the darker intergranular regions.

How can a layer become convectively unstable? The Schwarzschild criterion describes whether energy transport occurs by convection or radiation. We assume a gas bubble that is moving upward. The surrounding gas becomes cooler at a greater distance from the solar center; the cooling can be described by the radiative temperature gradient. The variation of the temperature of the gas bubble can be described as an adiabatic process since it is moving upward at high speed and no exchange of heat occurs. Therefore, the two temperature gradients become important:

- Gas bubble: adiabatic temperature gradient

$$\left|\frac{dT}{dr}\right|_{ad}. \tag{2.15}$$

- Surrounding gas: radiative temperature gradient

$$\left|\frac{dT}{dr}\right|_{rad}. \tag{2.16}$$

If the adiabatic gradient is smaller than the radiative gradient, the upward motion continues, convection occurs. Therefore, the Schwarzschild criterion for the onset of convection becomes

$$\left|\frac{dT}{dr}\right|_{ad} < \left|\frac{dT}{dr}\right|_{rad}. \tag{2.17}$$

We can expect convection in the two extreme cases:

- very small adiabatic gradient,
- very large radiative gradient.

The radiative gradient is

$$\frac{dT}{dr} = -\frac{3\kappa\rho L(r)}{4acT^3 4\pi r^2}, \tag{2.18}$$

where κ is the Rosseland mean opacity, $a = 7.56 \times 10^{-16}\,\mathrm{Jm^{-3}\,K^{-4}}$ is the radiation constant and $L(r)$ the luminosity at radius r. The net radiative energy flux is given by

$$F_{rad} = \frac{16\sigma T^4}{3\kappa}\frac{dT}{dr}, \tag{2.19}$$

where $\sigma = 5.67 \times 10^{-8}\,\mathrm{Wm^{-2}\,K^{-4}}$ is the Stefan–Boltzmann constant.

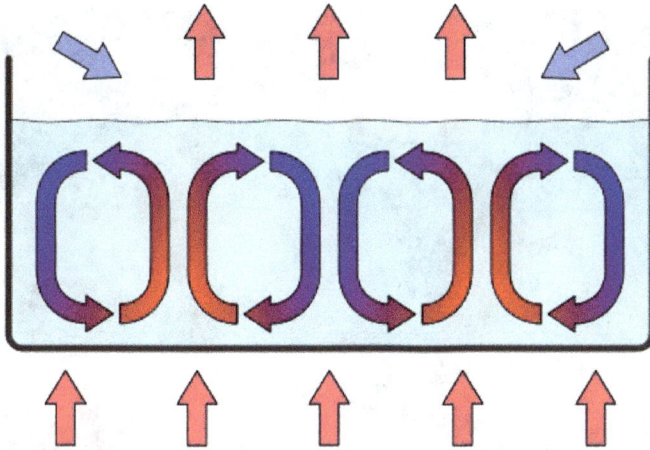

Fig. 2.2   Convection: principle. CC BY SA 3.0.

The adiabatic gradient can be written as

$$\frac{dT}{dr} = \frac{T}{P}\left(1 - \frac{1}{\gamma}\right)\frac{dP}{dr}. \qquad (2.20)$$

In Fig. 2.2 convection is shown in a simplified way. Heating occurs at the bottom and hot plasma or air rises, cools and finally sinks down. Then the process starts again.

One method for detecting planets around other stars is to observe the variation of the brightness of the star, when a planet passes in front of it. In this case, the granulation on a star has to be considered as an intrinsic uncertainty (stellar variability) on the precise measurements of exoplanet transits [Chiavassa *et al.*, 2017]. The full characterization of the granulation is essential for determining the degree of uncertainty on the planet parameters. In this context, the use of three-dimensional RHD simulations is important for measuring the convection-related fluctuations. This can be achieved by performing precise and continuous stellar photometry and radial velocity measurements, before, after and during the transit periods.

Figure 2.3 shows an artist's impression of the solar interior structure.

Acoustic pressure waves, P-modes, propagate in the Sun. These waves are generated by the turbulence in the convection zone, and certain frequencies are amplified. The oscillations are detectable as a Doppler shift pattern. Individual oscillations in the Sun are damped. They survive only a few periods. Interference between the localized waves produce global standing waves. These waves are known as normal modes. The oscillation

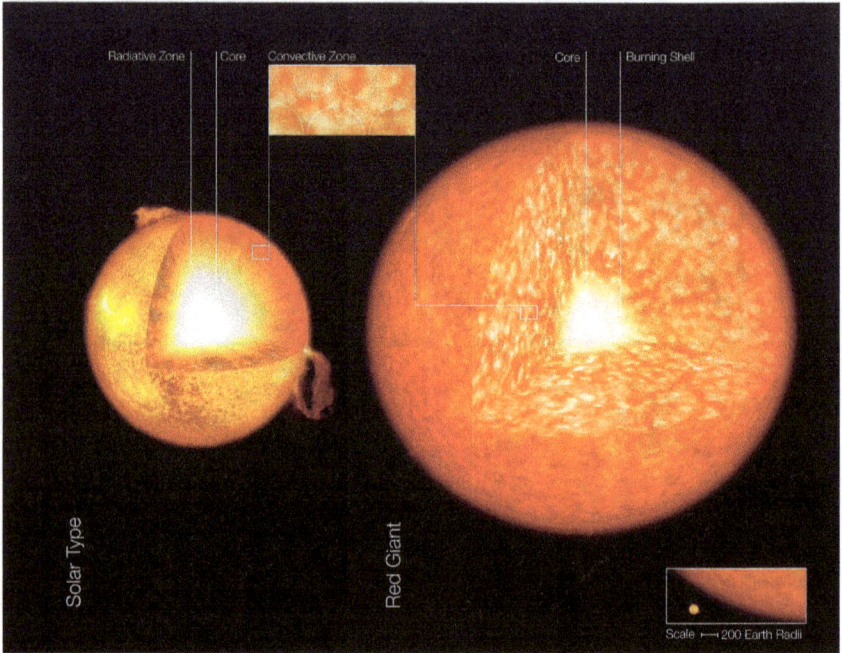

Fig. 2.3   Artist's impression of the solar interior. Left: present Sun; Right: Sun as a red giant. Credit: NASA.

modes are divided into three categories, and their classification depends on the restoring force:

- P-mode or acoustic waves: the restoring force is pressure. The variation of the speed of sound inside the sun is given by

$$v_s = \sqrt{\frac{\gamma P}{\rho}} = \sqrt{\frac{\gamma kT}{\mu m_H}} \approx 8.5 \times 10^7 \sqrt{\frac{\gamma T}{\mu}}, \qquad (2.21)$$

where μ is the mean molecular weight and m_H the mass of the hydrogen atom. Therefore, the sound speed increases with temperature and the temperature increases at deeper layers in the solar interior.
- P-modes occur at frequencies $> 1\,\text{mHz}$, are very strong in the range $2-4\,\text{mHz}$ and are often referred to as 5-min oscillations (5 min correspond to $3.33\,\text{mHz}$).
- G-modes or gravity waves: negative buoyancy, gravity acts as restoring force. The frequency of these modes is $<0.04\,\text{mHz}$. Being confined below

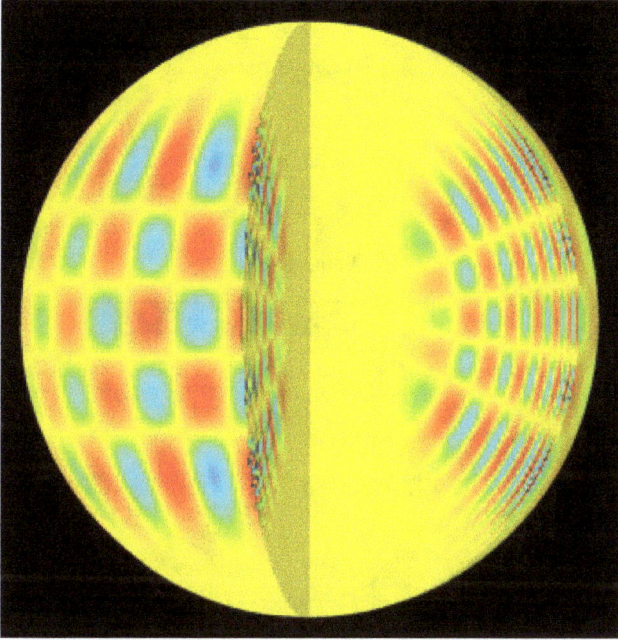

Fig. 2.4 A computer-generated image showing the pattern of a P-mode solar acoustic oscillation both in the interior and on the surface of the sun ($l = 20$, $m - 16$, and $n = 14$) Note that the increase in the speed of sound as waves approach the center of the sun causes a corresponding increase in the acoustic wavelength. Credit: NASA/SOHO.

the convection zone, these modes have residual amplitudes of a few millimeter in the photosphere and are extremely hard to detect.

- F-modes, surface gravity waves. They occur at or near the photosphere.

The oscillatory pattern in the solar interior can be seen in Fig. 2.4.

A recent review about helioseismology was given by Basu [2016]. From the analysis of the propagating waves, we can get information about different physical parameters in the solar interior, such as temperature, density or solar rotation.

2.2.4 *The solar atmosphere*

The solar atmosphere can be divided into

- photosphere,
- chromosphere,
- corona.

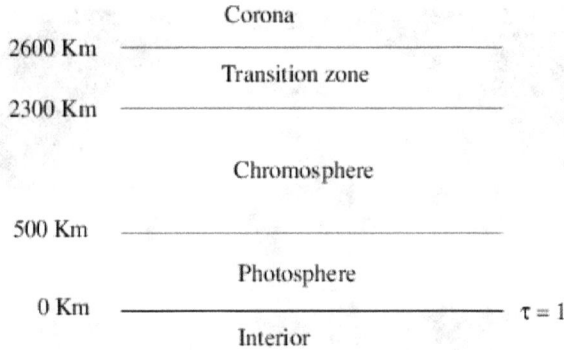

Fig. 2.5   The different layers of the solar atmosphere. τ denotes optical depth. At a depth of $\tau = 1$, the intensity is reduced by $1/e$.

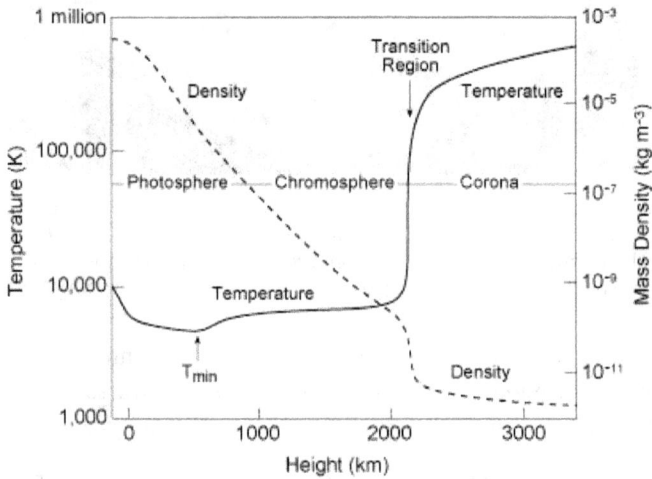

Fig. 2.6   Variation of temperature and density (dashed line) as a function of height in the solar atmosphere. Credit: NASA.

The extension of the different layers is shown in Fig. 2.5. In Fig. 2.6 the variation of temperature and density (dashed line) is shown as a function of height in the solar atmosphere.

Note that the temperature increases at larger heights. This is in contradiction with what we would expect, and therefore some heating mechanism is required to explain this.

The photosphere is the visible part of the solar atmosphere. It is very thin and extends about 400 km above the solar surface. The surface

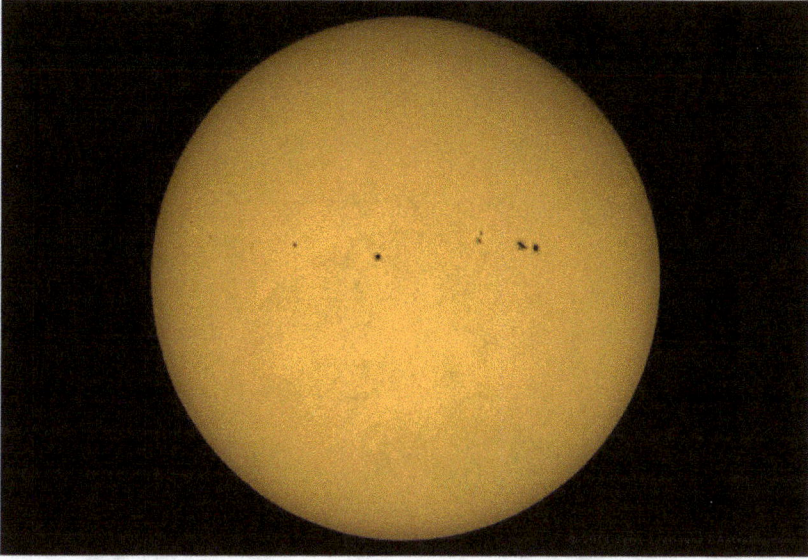

Fig. 2.7   White-light image of the Sun.

temperature is about 5,800 K. In Fig. 2.7 a so-called white-light image of the sun is shown. Near the solar disk center we see into deeper and hotter layers than near the limb, which causes a center-to-limb variation of intensity.

The solar chromosphere has an extension of several 1,000 km, and the temperature rises to several 10,000 K. It can only be observed during a total solar eclipse or in several spectral lines, most of which are in the UV, and therefore satellite observations are required. From ground-based observatories, the chromosphere can be observed in the light of Hydrogen Alpha. The so-called Hα-Line is formed by a transition of the electron in a hydrogen atom from energy level 3 to 2 (emission) or 2 to 3 (absorption). The central wavelength is about 656.3 nm. The core of this line is formed in the chromosphere, whereas the wings are formed in deeper layers. Additional chromospheric lines are the Ca II H- and K-lines. These lines occur near the UV at wavelengths 396.9 nm and 393.4 nm.

The solar corona can also be observed during a total solar eclipse (Fig. 2.8) or in spectral lines (in the UV or X-rays). In the transition zone, the temperature rises steeply from the chromosphere (several 10,000 K) to the corona (more than 1 million K). One can make a tomography of the solar atmosphere from the photosphere to the corona by observing the different spectral lines that are formed at these heights.

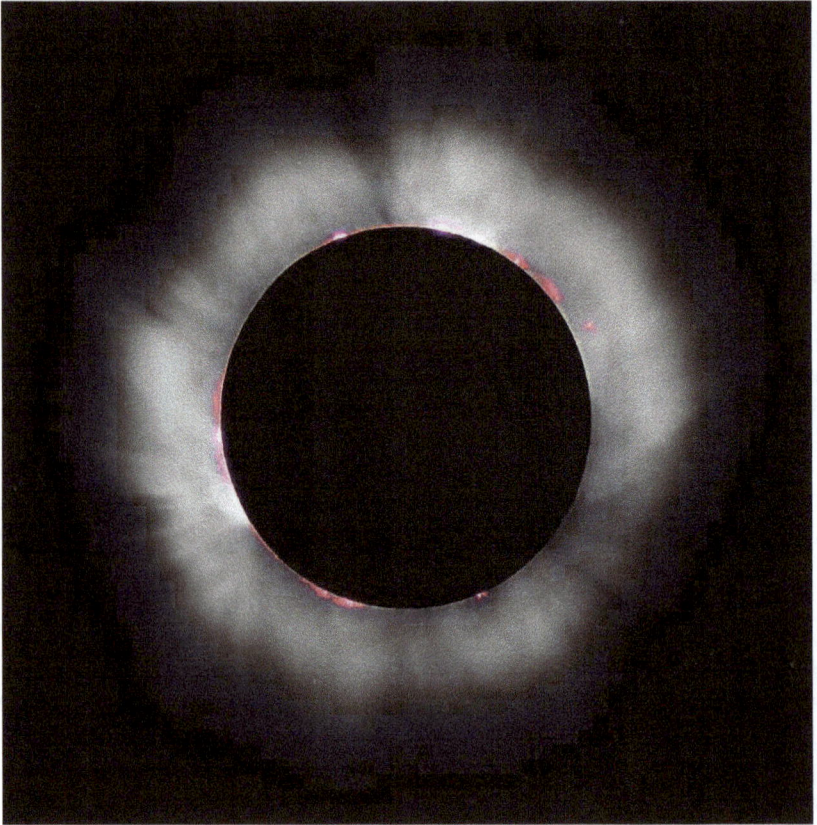

Fig. 2.8 The solar corona is visible during a total solar eclipse. Near the limb, one sees the red glowing chromosphere and some prominences. The image was taken during the total solar eclipse in 1999 which was visible in Europe. Credit: Luc Viatour/ www.Lucnix.be.

2.3 The active Sun

The Sun is the only star in which we can observe stellar activity in detail. The detection of stellar activity can only be made by indirect measurements.

2.3.1 *Sunspots*

Sunspots can even be seen with the naked eye as soon as they are extended by several 10,000 km. The earliest sunspot recordings were made in ancient China. The first telescopic observations of sunspots were made in the year 1609 by Galileo Galilei and others.

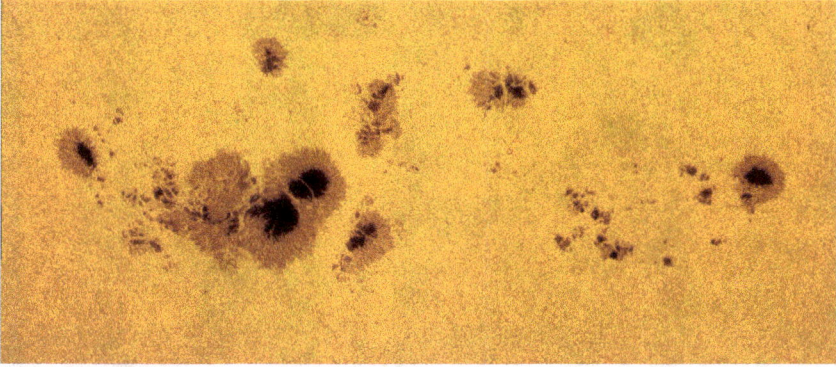

Fig. 2.9   Large sunspot groups. Credit: NASA/SOHO.

Sunspots appear dark because they have temperatures in the range of 3,000–4,500 K, which yields a high contrast with respect to the surrounding photosphere (5,780 K). Most sunspots consist of a dark, less structured and cooler umbra that is surrounded by a filamentary penumbra (see also Fig. 2.9).

The convective energy transport from the deeper, hotter layers is inhibited due to the presence of strong magnetic fields. Therefore, sunspots appear dark. The magnetic fields can be measured by using the Zeeman splitting of spectral lines, which is given by

$$\Delta\lambda = \lambda^2 Hg, \qquad (2.22)$$

where H is the magnetic field strength, g the Landé factor (can also be zero which means this line does not show any splitting). In Fig. 2.10 the splitting of spectral lines is shown when the spectrograph entrance slit (see image on the right) is put over a sunspot group.

In sunspots, the magnetic pressure $B^2/2\mu_0$, $\mu_0 = 4\pi \times 10^{-7}$ H/m adds to the gas pressure which is proportional to the temperature T. If P_i denotes the pressure inside a spot and P_e the pressure outside, the equilibrium condition is

$$P_i + \frac{B^2}{2\mu_0} = P_e. \qquad (2.23)$$

Therefore, less temperature is required inside a spot to establish equilibrium (assuming $P = NkT$).

The solar activity is measured by counting individual sunspots, f, and counting the number of sunspot groups, g (which, in most cases, are bipolar

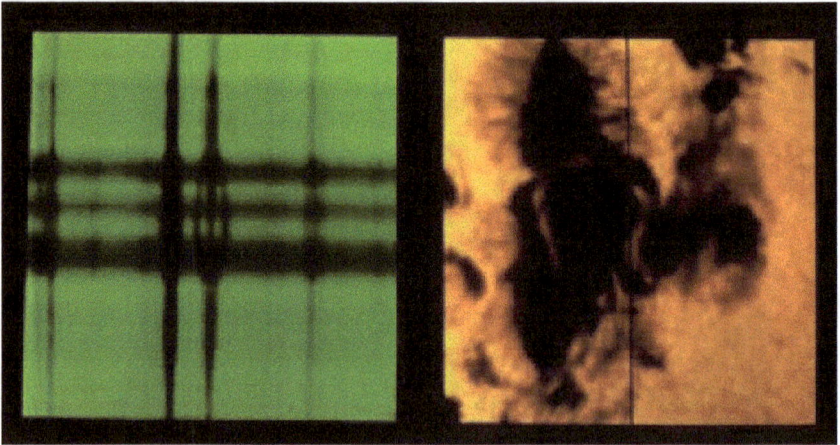

Fig. 2.10   Splitting of spectral lines merging from a sunspot group. Courtesy: NOAO.

groups). This gives the so-called Wolf number or sunspot index R:

$$R = k(10g + f), \tag{2.24}$$

where k is a factor that takes into account the optics of the telescope used and the seeing (stability of the Earth's atmosphere), both of which influence the number of detected spots.

After 17 years of making his own observations and comparing them with other data, S. H. Schwabe (1789–1875) discovered an 11-year periodicity in the number of sunspots. This was investigated further by R. Wolf (1816–1893). The sunspot number variation over the past 400 years, since the invention of the telescope, is shown in Fig. 2.11.

This figure shows some interesting peculiarities:

- the variation is not strictly 11 years;
- the amplitude of different cycles varies;
- there seem to be other periodicities than the so-called 11-year Schwabe cycle, e.g. the 80-year Gleissberg cycle or the 210-year Suess Cycle.
- between 1645 and 1715, practically no sunspots were observed. This is called the Maunder Minimum. The Dalton Minimum occurred around 1800 and is less pronounced. Between 1950 and 1990 the amplitudes were high, and this is referred to as the modern maximum.

There is not only just one periodicity of 11 years in the solar activity; intermittent phases, where solar activity was extremely low, are also found.

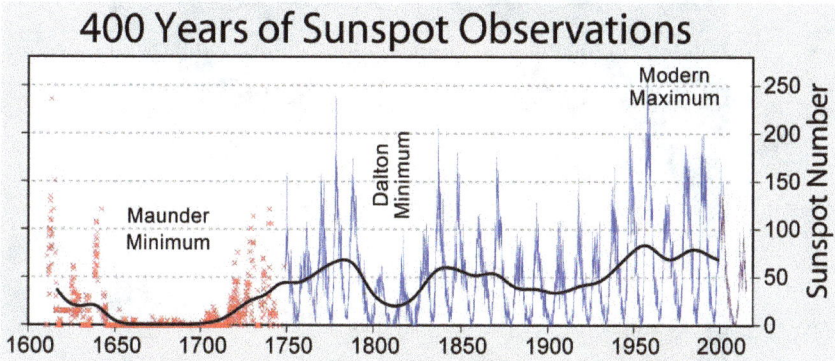

Fig. 2.11 Sunspot cycle. Since 1749, continuous monthly averages of sunspot activity have been available and are shown here as reported by the Solar Influences Data Analysis Center, World Data Center for the Sunspot Index, at the Royal Observatory of Belgium. Credit: Rhode, R.A., global warming project.

It seems to be established that we are entering a phase of low solar activity, and the next maxima could be of smaller amplitude. The sunspot cycles are numbered. The last cycle, number 24, began in 2008 and its maximum was less than the previous ones. The previous cycle, 23, lasted from 1996–2008. A summary of analyzing solar cycles was given by Solanki and Krivova [2011].

2.3.2 *Faculae*

Faculae are bright features; however, in the solar photosphere they can be only seen near the limb. Faculae always occur near sunspot regions. Like sunspots, faculae are also related to strong magnetic field concentrations. Because they appear bright, they are hotter than the surrounding photosphere and they overcompensate the radiation deficit that is caused by the dark sunspots. Therefore, the Sun becomes more luminous when there are many sunspots on its surface.

Faculae and sunspots contribute noticeably to variations in the "solar constant". The chromospheric counterpart of a facular region is called a plage.

In Fig. 2.12 solar faculae are shown near sunspots. This image was taken by NASA/SORCE and shows the chromosphere. In Fig. 2.7 a so-called white-light image of the Sun is seen, where faculae can be seen near the limb (note the limb darkening).

In the Hα-Line we can observe several chromospheric phenomena.

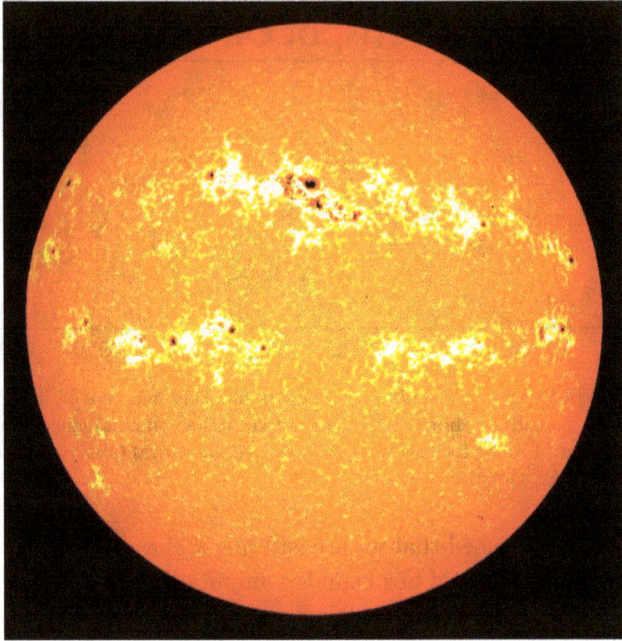

Fig. 2.12   Solar faculae near sunspots. Credit: NASA, SORCE.

2.3.3   *Prominences*

Prominences can be seen at the solar limb and are anchored to the Sun's surface in the photosphere, and extend outward into the Sun's corona. They consist of cooler plasma (several 100 times cooler) than the solar corona (more than 1 million K). In many cases, the plasma is formed into loops by magnetic fields. A prominence forms over timescales of about a day, and stable prominences may persist in the corona for several solar rotations, looping hundreds of thousands of kilometers into space (Fig. 2.13). If the magnetic field lines break up, plasma is expelled or falls back to the photosphere. Filaments appear as dark, strongly elongated structures on the solar disk and are the same as prominences but are seen across the solar disk (Fig. 2.14).

2.3.4   *Spicules*

Spicules are dynamic jets in the chromosphere, about 500 km in diameter, and are only seen at the solar limb. Matter moves upward of about 20 km/s;

Fig. 2.13 A huge solar prominence in comparison with Jupiter and Earth. Credit: NASA.

they last for about 15 minutes. It is estimated that about 300,000 spicules are distributed over the whole solar surface and they cover about 1% of the solar surface. If seen on the solar disk, they are also called fibrils or mottles (Fig. 2.15). They reach chromospheric heights of about 3,000–10,000 km. Spicules are also strongly associated with magnetic fields and, in these elements, the mass flux is about 100 times that of the normal solar wind.

2.3.5 *Flares*

A solar flare is a sudden brightening of a region mainly seen in the chromosphere. The brightening can be observed e.g. in the Hα-Line. The

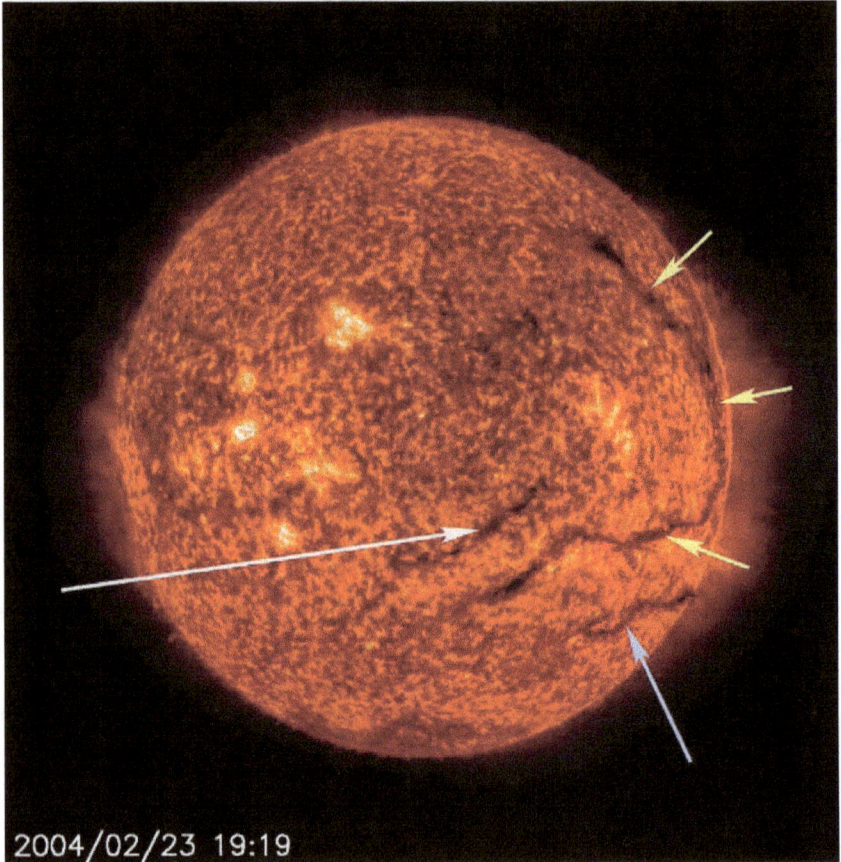

Fig. 2.14   Solar filaments seen on disk. Credit: NASA.

brightening is caused by a huge energy release corresponding to 10^{20}–10^{25} J within a few minutes. This energy corresponds to 1 billion megatons of TNT. For comparison, the energy that was released by the Hiroshima bomb was 15 kT TNT. Flares are eruptive processes in which clouds of electrons, protons and other heavier ions are accelerated near the speed of light. Also, radiation over the whole electromagnetic spectrum is produced, but mainly at short wavelengths. Flares are associated with active regions and complex sunspot groups that exhibit a complex magnetic structure, often a mixture of polarities. The X-ray and UV radiation emitted during flaring processes affect the Earth's ionosphere and can disturb radio communication. We will discuss these effects in Chapter 3.

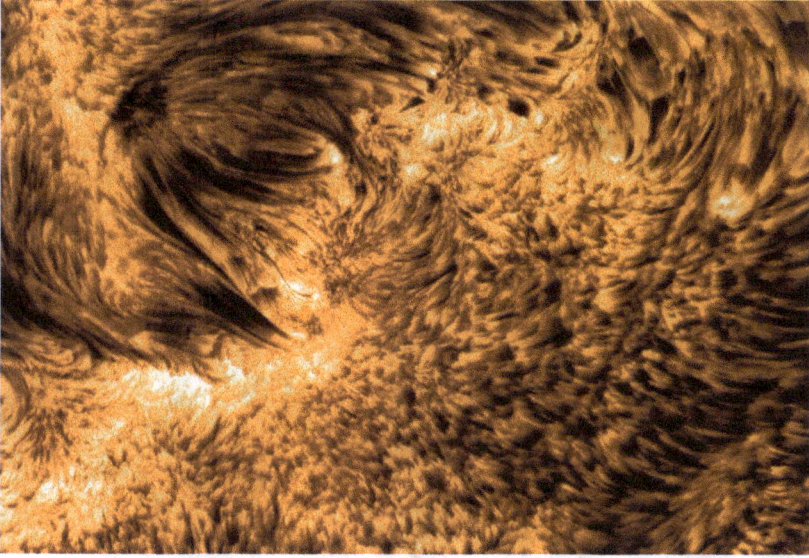

Fig. 2.15 Spicules, visible as dark tubes. Solar active region 10380, June 2004. Credit: SST, Royal Swedish Academy of Science.

The first flare was observed by R. C. Carrington and R. Hodgson in 1859. They independently observed a brightening within a small area in a big sunspot group. Flares that are visible in the photosphere are also called white-light flares. Flares are classified according to the peak flux in W/m^2 in the 0.1–0.4 nm range (X-rays):

(1) class A: $<10^{-7}$,
(2) class B: 10^{-7}–10^{-6},
(3) class C: 10^{-6}–10^{-5},
(4) class M: 10^{-5}–10^{-4},
(5) class X: $>10^{-4}$.

An X2 flare is twice as powerful as an X1 flare; X2 means $2 \times 10^{-4} \, W/m^2$ in the above-defined wavelength range. Another classification is based on the area of brightening in Hα. An example of an M9.9 class flare is given in Fig. 2.16.

2.3.6 *Coronal mass ejections (CMEs)*

A very spectacular phenomenon in the solar corona is coronal mass ejection, CME (Fig. 2.17). They are, sometimes, but not always, related

Fig. 2.16   Several wavelengths are combined in this New Year's Day solar flare image, categorized as an M9.9 and peaking at 1:52 p.m. EST on January 1, 2014. Each wavelength represents material at a different temperatures; from such observations we can study which processes move and heat the material. Credit: NASA/SDO.

to solar flares and are called solar storms in popular media. When a CME impacts the Earth's magnetosphere, it deforms the Earth's magnetic field and a geomagnetic storm is observed. CME impacts can induce magnetic reconnection in Earth's magnetotail (the midnight-side of the magnetosphere). This launches protons and electrons downward toward Earth's atmosphere, where they form the aurora. The average speed of a CME is about 500 km/s, but the speed can reach up to 3,000 km/s. The average mass ejected is 1.6×10^{12} kg. Their frequency of occurrence depends on the phase of the solar activity cycle:

• near solar minimum: about one every 5 days,
• near solar maximum: 3.5 per day.

2.3.7   *Coronal holes, solar wind*

The structure of the corona is dominated by magnetic fields that can be either closed or open (Fig. 2.18). In the coronal holes, the magnetic field

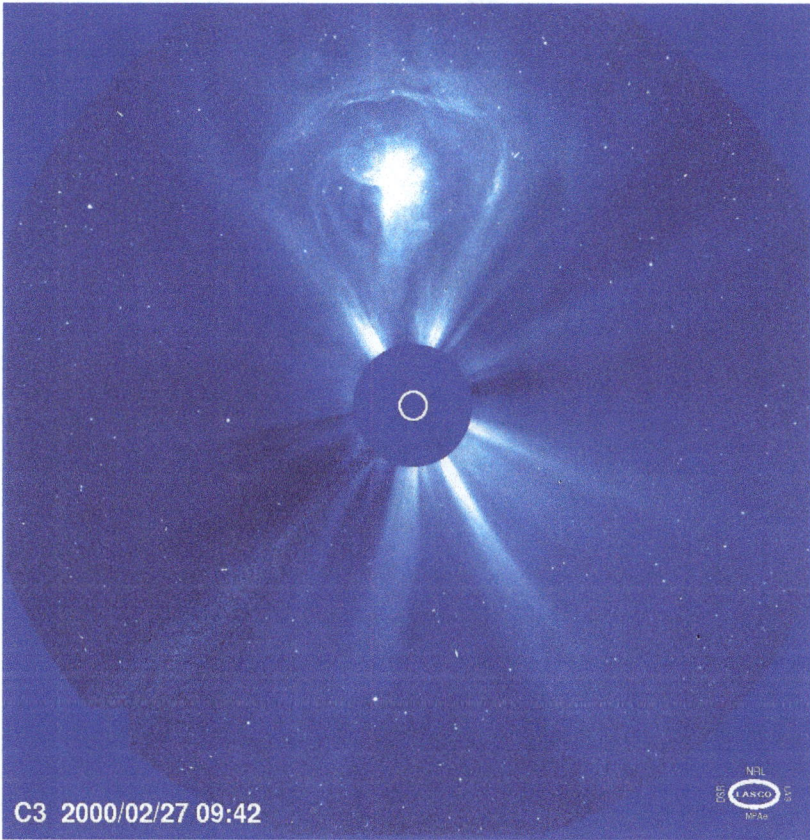

Fig. 2.17   Coronal mass ejection. SOHO mission. Credit: ESA/NASA.

appears unipolar (in fact, the field lines are closed very far away from the Sun in the interplanetary medium), and the fast component of the so-called solar wind is emitted through these open field lines. Coronal holes occur near the solar poles and appear dark in X-rays since less radiation is emitted. The solar wind consists of electrons, protons and ions with energies ranging from 1.5 keV to 10 keV. The solar wind consists of two components:

- fast speed component: typical velocity 750 km/s, temperature 8×10^5 K, composition matches the solar photosphere. Emerges from coronal holes.
- slow solar wind: 400 km/s, temperature 1.4–1.6×10^6 K, composition matches the solar corona. Emerges from a region around the solar equator, also called streamer belt.

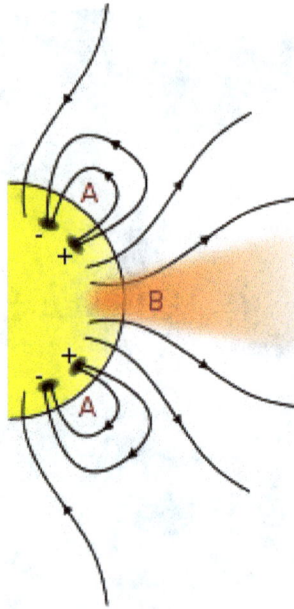

Fig. 2.18   Coronal arches connecting regions of opposite magnetic polarity (A) and the unipolar magnetic field in the coronal hole (B). Credit: Sebman81.

The solar wind pressure depends on the particle density n and the speed v:

$$P = 1.67 \times 10^{-6} n v^2, \tag{2.25}$$

where the pressure P is in nPa, n the number density in particles/cm^3 and the speed in km/s.

2.3.8   *The solar dynamo*

The solar dynamo theory explains the solar magnetic activity cycle, and such a theory can be also applied to model stellar activity cycles.

The Greek letter ω stands for angular velocity due to solar rotation. Let us consider magnetic fields inside the Sun. There, the conditions require that the field lines are driven by the motion of the plasma. Therefore, magnetic fields within the Sun are stretched out and wound around the Sun by differential rotation (the Sun rotates faster at the equator than near the poles). Let us consider a magnetic field line orientated north–south. Such a

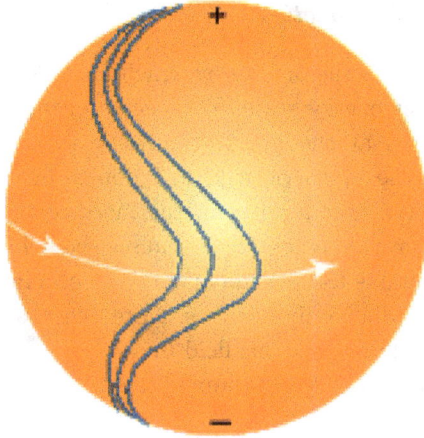

Fig. 2.19   Illustration of the ω-effect. The field lines are wrapped around because of the differential rotation of the Sun.

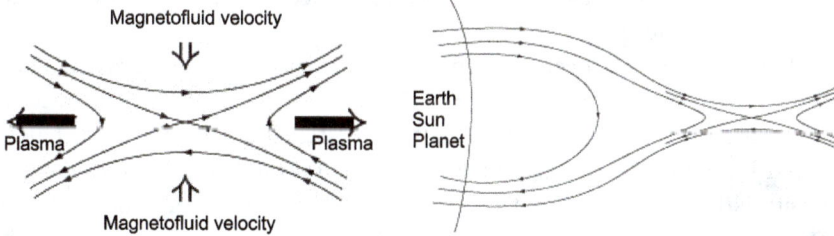

Fig. 2.20   Principle of magnetic reconnection.

field line will be wrapped around the Sun once in about 8 months because of the Sun's differential rotation (Fig. 2.19).

However, the field lines are not only wrapped around the Sun, but are also twisted by the Sun's rotation. This effect is caused by the coriolis force. Since the field lines become twisted loops, this effect is called α-effect. Recent dynamo models assume that the twisting is due to the effect of the Sun's rotation on rising flux tubes. These flux tubes are produced deep within the Sun.

A review on the features of solar and stellar dynamos is given by Charbonneau [2016].

2.3.9   *Magnetic reconnection*

Magnetic reconnection is the process by which lines of magnetic force break and rejoin in a lower energy state. The excess energy appears as kinetic energy of the plasma at the point of reconnection. In Fig. 2.20 single arrow lines denote the magnetic field and double-line arrows denote the magnetofluid velocity. As can be seen, the merging of two magnetofluids with oppositely oriented magnetic fields causes the field to annihilate. The excess energy accelerates the plasma out of the reconnection region in the direction of the full double-line arrows. Note the characteristic X-point, where the topology changes for the field lines.

The plasma, where the field is annihilated, is accelerated outward to Alfvén speed v_A:

$$v_A = B_0/\sqrt{4\pi M n_B}, \qquad (2.26)$$

where n_B is the density inside the current sheet, M the plasma average molecular weight.

A similar process occurs in coronal loops that were observed in hard and soft X-rays by Yohkoh and SOHO instruments. Such a coronal loop (see drawing on the right in Fig. 2.20) is stretched out by pressure that is caused by buoyancy. A magnetic structure is buoyant because the particle density is lower there since it contains larger magnetic energy density (see magnetic buoyancy). Thus, the external pressure is balanced by a lower gas pressure in conjunction with a magnetic pressure. The top of the loop distends and reconnection occurs. Particles in the reconnection region accelerate toward the surface of the Sun and out away. The particles that are accelerated toward the Sun are confined within the loop's magnetic field lines and follow these lines to the footpoint of the loop, where they collide with other particles and lose their energy through X-ray emissions. Such processes are the cause of solar flares and will be discussed in Chapter 3.

Magnetic reconnection also provides a mechanism for energy to be transported into the solar corona. Furthermore, reconnection occurs in the earth's magnetotail. The solar wind distends the Earth's dipole field so that the field extends far behind the Earth. Earthward flowing plasma streams with flow velocities up to 1,000 km/s (which is close to the local Alfvén speed) have been observed [Birn *et al.*, 1981].

A recent review on solar MHD was given by Walsh [2001].

2.4 Earth and space weather

2.4.1 *Definition of space weather*

Modern society has become strongly reliant on technologically advanced systems, often located in space, such as telecommunication, and navigation. Therefore, the question arises as to whether such systems are affected by the conditions and variations in space where these satellites orbit the Earth.

It is generally accepted that the term *space weather* refers to the time-variable conditions in the space environment that may effect space-borne or ground-based technological systems.

During the last few years, space weather activities have expanded worldwide. Examples for such activities, which are of national and international interest, are

- US Space Weather Program,
- US-NASA's Living With a Star program,
- ESA's space weather program,
- SWENET, Space Weather European Network,
- SIDC, Solar influences data center at the Royal Observatory in Belgium,
- Lund space weather center,
- The Australian IPS Radio and Space Services, the Australian Space Weather Agency,
- The Canadian Space weather program,

and many others (such as the group in Oulu, Finland). Today, space weather is monitored from a worldwide net of ground stations and from space. Both types of observations are complementary.

According to the US National Space Weather Programme, the definition of space weather is: *Conditions on the Sun and in the solar wind, magnetosphere, ionosphere and thermosphere that can influence the performance and reliability of space-borne and ground-based technological systems and can endanger human life or health.*

An artist's concept of the various space weather effects on Earth and space near Earth is shown in Fig. 2.21.

Space weather can affect:

- On Earth: radio communication, oil pipelines, power grids.
- Earth's atmosphere: heating, ionization.
- In space: electric charging of satellites, damage solar cells, be a danger for astronauts.

Fig. 2.21   An artist's concept of the different space weather effects on Earth and space near Earth. Credit: NASA.

2.4.2   *Examples of space weather effects in space*

Let us give some examples of space weather influencing on satellites[2]

- Space Shuttle: Numerous micrometeoroid/debris impacts have been reported.

---

[2]See also: Tribble, A. C. (2003). *The Space Environment*, Princeton Univ. Press, Princeton, NJ, USA.

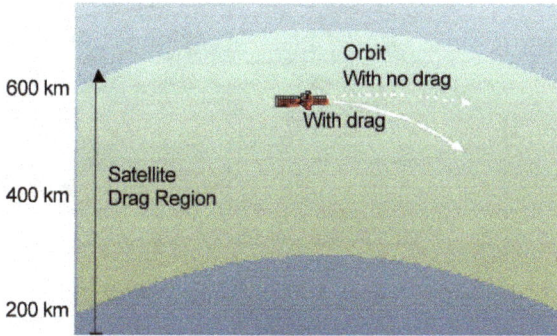

Fig. 2.22   Enhanced drag on satellites during solar flares and CMEs.

- Ulysses: failed during the peak of the Perseid meteoroid shower.
- Pioneer Venus: Several command memory anomalies were related during high-energy cosmic rays.
- GPS: Photochemically deposited contamination on solar arrays.

Short wavelength radiation from the Sun heats the Earth's upper atmosphere. Therefore the upper atmosphere expands and the drag on low Earth orbiting satellites is enhanced. This is illustrated by Fig. 2.22.

2.4.3   *Space weather and telecommunication*

On Earth, we could experience radio fade outs. The HF communication depends on the reflection of signals in the Earth's upper atmosphere. These layers are strongly influenced by the Sun's shortwave radiation.

For telecommunication, the ionosphere is extremely important since it bends, reflects or even absorbs radio waves. The so-called plasma frequency is given by

$$\nu_{p,e} = 9 \times \sqrt{n_e}, \tag{2.27}$$

where n_e is the electron density per m^3 and ν_c is given in Hz. This density depends on the degree of ionization and becomes enhanced during large solar CMEs and flares.

Let us first consider the Earth's ionosphere. The degree of ionization and the formation of different layers depends on the solar radiation, and there a clear day/night variation occurs (Fig. 2.23). At night, the E layer and F layer are present. During the day, a D layer forms and the E and F layers become much stronger. Often, the F layer will differentiate into

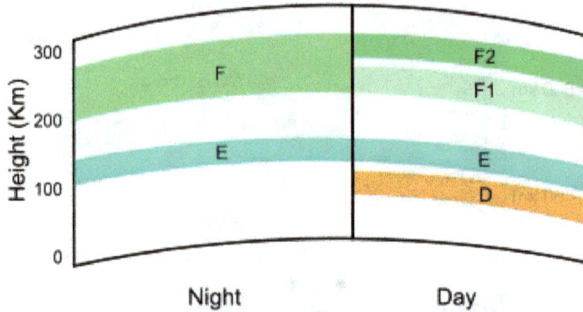

Fig. 2.23   Day/night variation of the Earth's ionosphere.

F1 and F2 layers during the day. Typical perturbations of the ionosphere related to solar activity are:

- SID, sudden ionospheric disturbances: A strong solar flare produces hard X-rays that hit the sunlit side of the Earth. These rays penetrate the D-region and the number of ionized atoms is increased there, which means the electron density is also increased. A high-frequency fade out (3–30 MHz) occurs because of signal absorption. Very low-frequency signals (3–30 kHz) are reflected in the D-layer instead of the E-layer. As soon as the X-rays ends, the atoms recombine and the normal state is established.

- Polar Cap Absorption, PCA: During large solar flares, protons are also emitted. They hit the Earth within 15 minutes to 2 h after the onset of the flare. The protons spiral around and down the magnetic field lines of the Earth and penetrate into the atmosphere near the magnetic poles, increasing the ionization of the D and E layers. PCA's typically last anywhere from about an hour to several days, with an average time span of around 24–36 h.

- Geomagnetic storm: This is a temporary, intense disturbance of the Earth's magnetosphere. During a geomagnetic storm, the F_2 layer becomes unstable and may even disappear. This causes beautiful aurorae in the Northern and Southern pole regions which can be observed.

2.4.4   *Examples of space weather events on Earth*

In 1989 (March 13), solar activity induced a huge geomagnetic storm, causing a saturation in the transformers and completely shutting down the

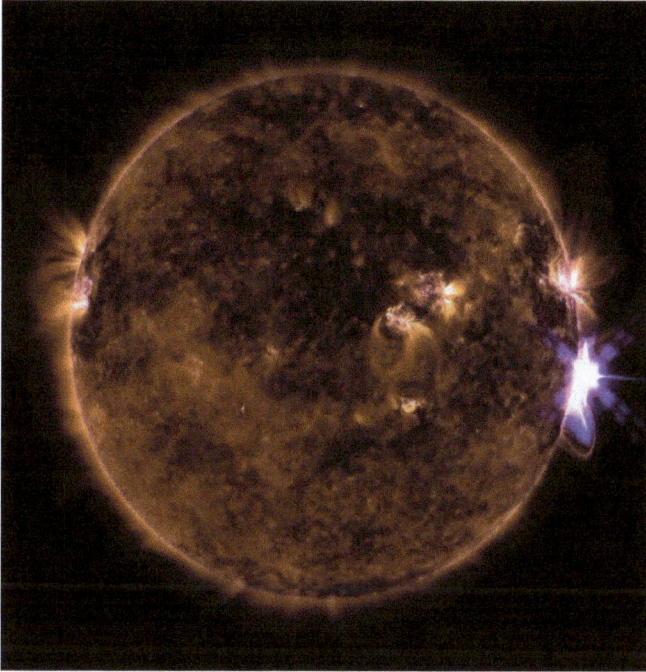

Fig. 2.24 A strong flare that occurred near solar minimum in September 2017. Credit: NASA.

power grid servicing Canada's Quebec province. The blackout resulted in a loss of 19.400 MW in Quebec and 1325 MW of exports. Service restoration took over 9 h (after R. Thompson, IPS, Radio and Space Service).

A surprising event occurred on September 10, 2017. The flare is shown in Fig. 2.24, the image shows a combination of wavelengths of extreme ultraviolet light that highlights the extremely hot material in flares, which is then colorized. It was classified as an X8.2 flare. The Sun was near the minimum of activity. HF radio communication was affected, a loss of contact for up to an hour over the sunlit side of Earth occurred and low-frequency communication degraded for about an hour.

Chapter 3

Exoplanets

In this chapter we give an introduction to the rapidly growing field of exoplanets, and we will start with the discussion how exoplanets can be detected.

The main problem in finding exoplanets comes from:

- Planets are smaller than stars.
- Planets only reflect light from a star.
- Seen from a large distance, they appear very close to their parent star.
- Planets mainly radiate in the IR.

The main methods are through observations of transits, radial velocity measurements, astrometric measurements, direct imaging and microlensing. Relevant satellite missions such as Kepler, COROT and GAIA will be mentioned. It is even possible to detect exoplanetary atmospheres. We will review the different types of detected exoplanets and, finally, we discuss aspects of the orbits of exoplanets and their stability.

3.1 Methods to detect exoplanets

3.1.1 *Transits*

If the orbital plane of an exoplanet lies in the same/similar plane as the one from an observer on Earth, we can see the planet passing in front of its host star (transiting exoplanet). This causes a periodic dimming of the stellar lightcurve and can be revealed by precise photometry. Figure 3.1 shows the lightcurve variations due to a transiting planet.

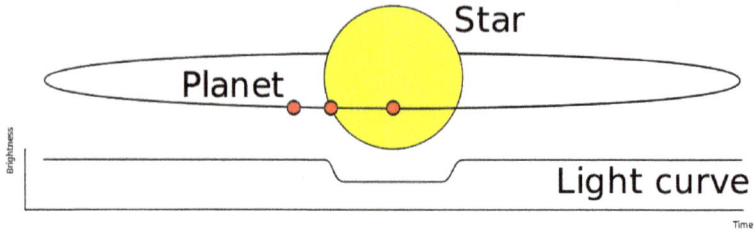

Fig. 3.1 The transit of a planet can be observed as a dip in the lightcurve of the star. Credit: Wikipedia Commons, CC BY-SA 3.0.

Table 3.1. Transit properties of solar system planets.

Planet	P (yr)	a (AU)	T (h)	D (%)	Prob. (%)	Incl. (deg)
Mercury	0.24	0.39	8.1	0.0012	1.19	6.33
Venus	0.62	0.72	11.0	0.0076	0.65	2.16
Earth	1.0	1.0	13.0	0.0084	0.47	1.65
Mars	1.88	1.52	16.0	0.0024	0.31	1.71
Jupiter	11.86	5.2	29.6	1.01	0.089	0.39
Saturn	29.5	9.5	40.1	0.75	0.049	0.87
Uranus	84.0	19.2	57.0	0.135	0.024	1.09
Neptune	164.8	30.1	71.3	0.127	0.015	0.72

For a circular orbit, the duration of a transit can be estimated from

$$\tau_{\mathrm{tr}} = 13d_* \sqrt{a/M_*} \sim 13\sqrt{a} \qquad \text{(h)}. \qquad (3.1)$$

a is the semi-major axis of the planet's orbit, M_* the mass of the star and d_* the stellar diameter in solar units. The duration alone does not give any information about the physical nature of the planet. The size of a planet follows from the transit depth because the fractional change in brightness is equal to the ratio of the planet's area to the star's area. As an example we give the transit properties of solar system objects in Table 3.1 (from http://kepler.nasa.gov/sci/basis/character.html). P is the orbital period in years, a the semi-major axis in AU, T the transit duration in hours, D the transit depth in %, "Prob." the geometric probability in % and "Incl." the inclination invariant plane in degrees.

A greater precision in photometric observations is achievable above the Earth's atmosphere from satellite missions, which will help to detect planets as small as Earth. One advantage of this method is that larger planets detected from photometric lightcurves can be tested and verified with the radial velocity method since this method works best under $i = 90°$. The precision of the measurements must be high. For example, in the case of HD 209458, the drop in the lightcurve is only 1.7%.

Exoplanets seen from Earth display phases like the Moon. When they transit the host star, the situation becomes similar to the New Moon; shortly before vanishing behind the host star or after reappearing from it, they become nearly fully illuminated, which resembles the Full Moon. This causes an additional small light variation of the system.

However, there are also other sources of light variation of a star:

- starspots,
- intrinsic stellar variability.

As starspots are rotationally modulated, one must be careful in the interpretation of lightcurve variations.

3.1.2 *COROT and Kepler*

Two important satellite missions for the detection of exoplanets are the Kepler mission and the COnvection ROtation and planetary Transits (COROT) mission. Both missions focus on the detection of exoplanets by measuring the brightness variations of their host stars (transit method).

COROT was launched in December 2006. Due to a computer failure, the transfer of data stopped in 2012. The telescope's diameter was 27 cm and 4 CCDs (2048 × 2048 pixel) were used for the measurements. In order to avoid the Sun entering in its field of view, COROT observed an area around Serpens Cauda, toward the galactic center during the northern summer, and it observed in Monoceros, in the Galactic anticenter during the winter.

The exoplanet research program requires a large number of dwarf stars to be monitored, and the avoidance of giant stars, for which planetary transits are too shallow to be detectable. COROT also executed an asteroseismic program, which requires stars brighter than magnitude 9, and many different stellar fields need to be covered. In order to confirm the observation, two transits should be observed. Candidates that show only one transit have been found, but uncertainty remains about their exact orbital period.

COROT-1b was the first planet detected by COROT, and it is a hot Jupiter. By further analysis, COROT-1b became the first exoplanet to have its secondary eclipse detected in the optical range, thanks to the high-precision lightcurve delivered by COROT. Another interesting planet is COROT-3b; it has a mass of 22 $M_{\rm Jup}$. Therefore, it appears to be something

Fig. 3.2 The array of CCD cameras used on the Kepler satellite mission to detect exoplanets by photometry. Credit: NASA.

between a brown dwarf[1] and a planet. More than 600 exoplanets have been detected with COROT.

The Kepler mission continuously monitors about 150,000 stars. This mission was launched in 2009. On January 6, 2015, NASA announced the 1000th confirmed exoplanet discovered by the Kepler Space Telescope. The telescope consists of a Schmidt camera with a 0.95-m front corrector plate (lens) feeding a 1.4-m (55 in) primary mirror. The Schmidt camera contained in total more than 40 CCD cameras (Fig. 3.2).

The Kepler search space in the Galaxy is shown in Fig. 3.3. A region extending to about 3000 light years was examined.

3.1.3 *Microlensing*

Gravitational microlensing occurs when the gravitational field of a star acts like a lens, magnifying the light of a distant background star. If the foreground lensing star has a planet, then the planet's own gravitational field can make a detectable contribution to the lensing effect of the star. This can be observed in the variation of the lightcurve as a secondary bump (see Fig. 4.12).

---

[1] Brown dwarfs are substellar objects that occupy the mass range between the heaviest gas giant planets and the lightest stars.

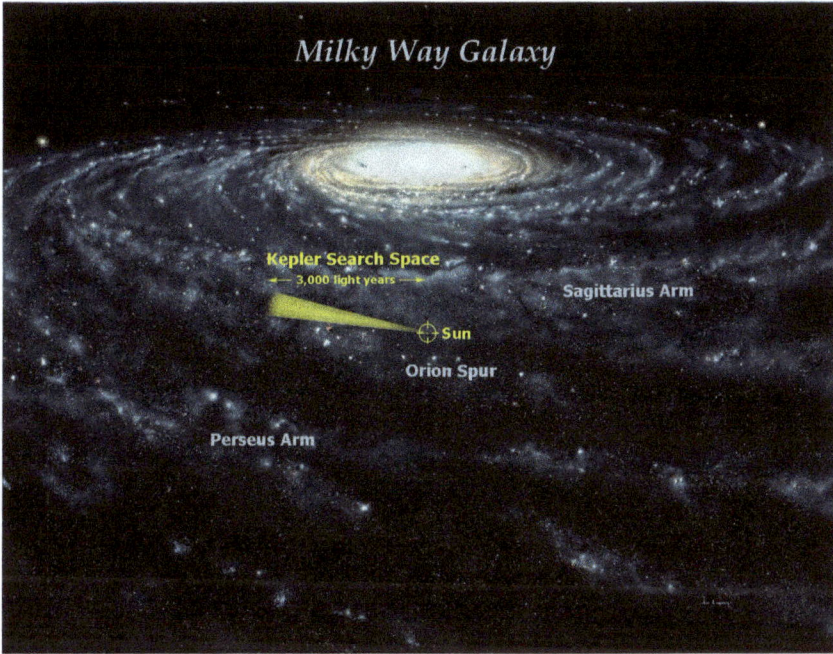

Fig. 3.3 The location of the Sun in the galaxy and the search space of the Kepler mission. Credit: NASA.

D_L denotes the distance between the observer O and the lens, D_{LS} is the distance between the lensing star and the source S, and D_S is the distance between the observer and the source. An amplification of the light due to microlensing can be observed when the object moves within the Einstein radius defined by

$$\Theta_E = \sqrt{\frac{4GM}{c^2}\frac{D_{LS}}{D_L D_S}}. \tag{3.2}$$

Under favorable circumstances, planets as small as Earth could be detected.

The main disadvantage of this method is that lensing cannot be repeated since the alignment never occurs again. The detected planets will be several kpc away, thereby making the other methods inapplicable for confirming the observation. On the other hand, through a systematic observation we would be able to get the basic information about the number of Earth-like planets in the galaxy.

Microlensing observations are made by robotic telescopes distributed worldwide. One network is the NASA/NSF-funded Optical Gravitational

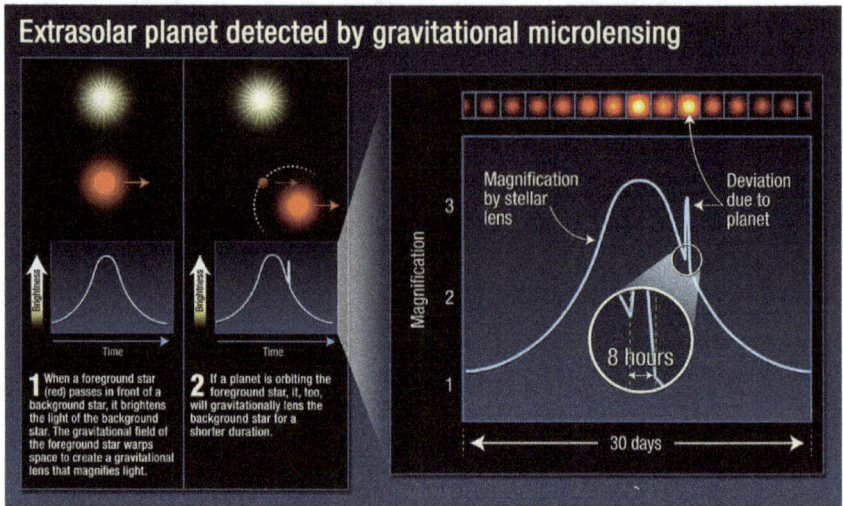

Fig. 3.4   Principle of microlensing. Credit: NASA.

Lensing Experiment (OGLE), another project is Probing Lensing Anomalies NETwork (PLANET). The instrument is a 1.3-m telescope.

One aim of OGLE is to detect exoplanets, another aim is to detect dark matter by gravitational lensing effects. Therefore, the targets in the sky are the small and large Magellanic clouds and the galactic bulge. In both cases, a huge number of background stars can be seen, thus the probability for the detection of a microlensing event is enhanced. The first experiments started in 1992.

The planet OGLE-2005-BLG-390Lb is known as a super-Earth planet, located at a distance of 21,500 ly from the solar system. The host star has spectral type M4, the mass of the planet is 5.5 Earth masses and its distance to the host star is about 2.6 AU. Since the host star is a cool M star, the surface temperature expected on the planet is only 50 K. It was detected by the Danish 1.54 m telescope at ESO, a telescope which is part of the PLANET network [Dominik *et al.*, 2002].

A general overview on microlensing detection method was given by Bennett [2008].

3.1.4 *Direct observations*

Direct observation of exoplanets means imaging. Extrasolar planets are very faint objects located near much brighter stars, so they are extremely difficult to observe directly. The light from the star and the planet has to be separated. The first planet discovered by this technique orbits a brown dwarf. In that case, the contrast between star and planet was less.

Direct imaging becomes extremely important when trying to analyze chemical processes in the atmospheres of exoplanets (see also Moses *et al.* [2016]).

Only ~3% of the currently confirmed exoplanets have been detected through direct imaging. Generally, this method works best for young planets that emit infrared light and are far from the glare of the star.

In 2008, three new planetary systems orbiting A-type main sequence stars were announced. The star HR8799 seems to contain four directly imaged planets. The planets orbiting this star have wide orbits (larger than 25 AU) and are very young and very massive [Marois *et al.*, 2010].

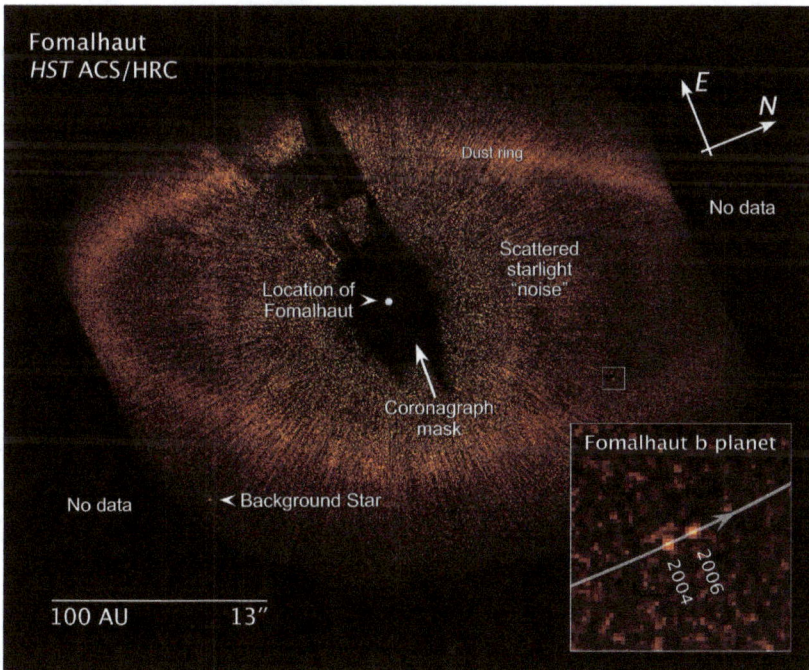

Fig. 3.5   This annotated image shows key features of the Fomalhaut system, including the newly discovered planet Fomalhaut b (aka Dagon), and the dust ring. Also included are a distance scale and an insert, showing how the planet has moved around its parent star over the course of 21 months. The Fomalhaut system is located approximately 25 light years from the Earth. Credit: NASA, ESA, and Z. Levay (STScI).

The star Fomalhaut (αPsa) (see also Fig. 3.5) is at a distance of 7.7 pc from Earth. Its spectrum shows a strong IR excess, indicating a circumstellar disk. The inner edge of the toroidal debris disk is found at a distance of 133 AU from the star, the belt having an extension of 25 AU. In 2008, a planet just orbiting at the inner edge of the debris disk was found using Hubble Space Telescope (HST) observations. The mass determination is not definite yet, but the planet's mass is somewhere between half the mass of Neptune and three times the mass of Jupiter. Observations of the star's dust ring by the Atacama Large Millimeter Array point to the existence of two planets in the system, neither one at the orbital radius proposed for the HST-discovered Fomalhaut b [Boley et al., 2012].

Chauvin et al. [2005] report deep imaging observations of the young, nearby star AB Pic. They detected a faint, red source $5.5''$ south of the star with JHK colors compatible with that of a young substellar L dwarf. Follow-up observations at two additional epochs confirmed that the faint red object is a companion to AB Pic rather than being a stationary background object. A low-resolution K-band spectrum indicates an early L spectral type for companion. Finally, evolutionary model predictions based on the JHK photometry of AB Pic b indicate a mass of 13–14 M_J if its age is $\sim$30 Myr.

3.1.5 Pulsar timing

Stars with masses larger than 1.4 solar masses explode into a supernova. The explosion starts with an implosion; but the inner mass contracts, while the outer parts are expelled. The final stage of stellar evolution of stars in that mass range is a neutron star, whereas stars with masses larger than about 4 solar masses end up as black holes.

Stars generate their energy by thermonuclear fusion reactions. These reactions end with the formation of iron. Therefore, an iron core develops. As soon as the iron core exceeds 1.4 $M_\odot$ it collapses to a size of about 10 km. Supernovae eject newly formed massive elements into interstellar space. Neutron stars have enormously strong magnetic fields. Electrons and positrons moving in the neutron star's magnetic field produce radiation. This radiation is beamed away from the poles of the neutron star. As the neutron star rotates, these beams sweep around like the beams of a lighthouse. As the beam sweeps past an observer, the neutron star appears to pulse on and off. It is, therefore, called pulsar.

The intrinsic rotation of a star is quite regular. The slightest variations in the pulses caused by the pulsar's motion can be easily measured. During orbiting of a planet, the pulsar and the planet move around the center of gravity. These tiny motions can be measured so precisely from the pulses, such that masses down to 1/10 Earth masses can be detected. In 1992, A. Wolszczan and D. Frail used this method to discover planets around the pulsar PSR 1257+12. This was, in fact, the first confirmation of the detection of extrasolar planets.

Planets around pulsars and other evolved stars were discussed in the book of Wolszczan and Kuchner [2010]. So far, eight circumbinary planets (these are planets orbiting a binary star) have been discovered in this way.

3.1.6 *Radial velocities*

We have seen how astrometric measurements can be used to determine the motion of a star around the center of gravity due to the gravitational influence of the planet. Of course, such a motion also leads to variations in the speed at which the star moves relative to Earth. By using precise Doppler measurements of spectral lines in the spectrum, displacements of spectral lines can be detected (see Fig. 3.6). Since the star is more

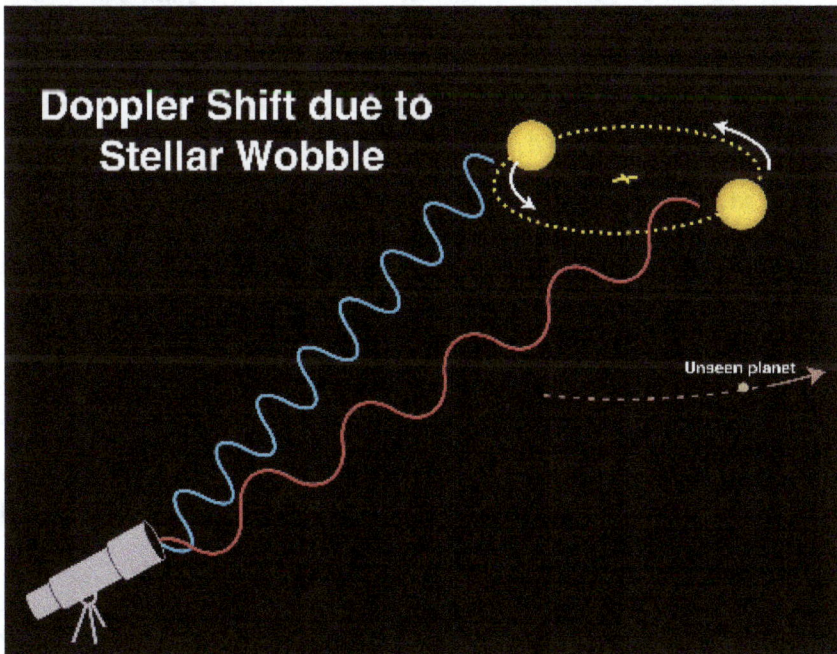

Fig. 3.6 The motion of the central star about the barycenter due to the presence of an exoplanet can be measured by a periodic Doppler shift of spectral lines. Credit: astrobio.net.

massive than the planet, these motions are extremely small and precision of measurements down to 1 m/s is required.

The Doppler velocity can be derived from the Doppler formula:[2]

$$\lambda' = \lambda_0 (1 + v/c). \tag{3.3}$$

λ' is the wavelength of the line coming from a source moving with velocity v in radial direction relative to an observer, λ_0 is the velocity of the spectral line when the source is at rest relative to an observer and c is the speed of light. This formula is valid only for non-relativistic speeds, i.e. $v \ll c$.

Such Doppler shifts are measured with spectrometers. Early spectroscopes were made of prisms, but modern spectroscopes use a diffraction grating. Two examples of modern high-resolution spectrometers are the high accuracy radial velocity planet searcher (HARPS) at the ESO 3.6 m telescope in La Silla Observatory and the HIRES at the Keck Telescope.

The method is very precise but requires a high signal-to-noise ratio and therefore works only for objects closer than 160 ly. Massive planets close to their host stars can easily be found (because of the short period of revolution around the host star); however, the detection of massive planets at greater distances requires years of high-precision observation.

Let us consider the case of our planetary system. For example, consider Jupiter. The duration of its orbital period is about 12 years. Therefore, since Jupiter contains about 75% of the mass of all other planets, the velocity signal of the motion of the Sun with respect to the barycenter of the solar system will have a period of roughly 12 years.

Planets with orbits highly inclined to the line of sight from Earth produce smaller wobbles and are more difficult to detect.

The semi-amplitude, K, of the radial velocity of a star of mass M_* that is induced by an orbiting planet of mass M_P is

$$K = \left(\frac{2\pi G}{P_{\rm orb}} \right)^{1/3} \times \frac{M_P \sin i}{(M_* + M_P)^{2/3}} \frac{1}{\sqrt{1 - e^2}}, \tag{3.4}$$

where $P_{\rm orb}$ is the orbital period, i the angle between the normal to the orbital plane and the line of sight and e the eccentricity of the planet's orbit. We see that for $i = 0$ there is no amplitude since we cannot measure any radial velocity component.

The first planet discovered using the technique of radial velocity measurements was 51 Pegasi b Mayor and Queloz [1995]. The companion

---

[2]Ch. Doppler, 1803–1853.

lies only about 8 million km from the star, which would be well inside the orbit of Mercury in our solar system. This object might be a gas-giant planet that has migrated to this location through orbital evolution.

3.1.7 *Astrometry*

A planet orbiting a host star can be fully described as a two-body system. Both objects move around their barycenter which, in most cases, lies inside the star because of its mass m_1 is larger than the planet's mass m_2.

The center of the mass will be located near the more massive component. Let us consider the center of mass of the system Sun and Jupiter (the most massive planet in the solar system). Since the mass of the Sun is about 1,000 times the mass of Jupiter, the center of mass of such a two-body system must be located 1,000 times nearer to the Sun than to Jupiter.[3] Therefore, precise position measurements are required to detect the motion of a star around the center of gravity in case of the presence of other planets.

The motion of the barycenter of the solar system relative to the Sun is shown in Fig. 3.7.

The reflex motion of the host star due to the presence of a planet is, in many cases, the only means to derive accurate planetary and stellar masses.

The radial velocity method described in the next section yields true masses only in the case when transit observations are possible, i.e. when the system is seen nearly edge on.

The astrometric signal can be estimated from

$$\Theta_{max} = 1.91 \times \frac{a_{comp}}{[AU]} \times \frac{[pc]}{d} \times \frac{M_{comp}}{[M_J]} \times \frac{[M_\odot]}{M_*}, \qquad (3.5)$$

where a_{comp} is the semi-major axis, M_{comp}, the mass of the planetary companion orbiting a star with mass M_* at a distance d.

Let us consider an exoplanet with a mass of 1 M_J around a solar-mass star in a two years orbit at a distance of 10 pc. Then, from Kepler's third law

$$\frac{a^3}{P^2} = \frac{G}{4\pi^2}(M_* + M_{comp}), \qquad (3.6)$$

the astrometric signal will be 0.3 mas = 300μas.[4] A planet having a mass of 10 M_J at distance 100 pc orbiting a star with mass 0.5 $M_\odot$ in 1 year will induce an astrometric signal of about 300 μas.

---

[3]In fact, it is located just outside the Sun's sphere.
[4]mas means milliarcsec, 1 mas = $1/1000''$.

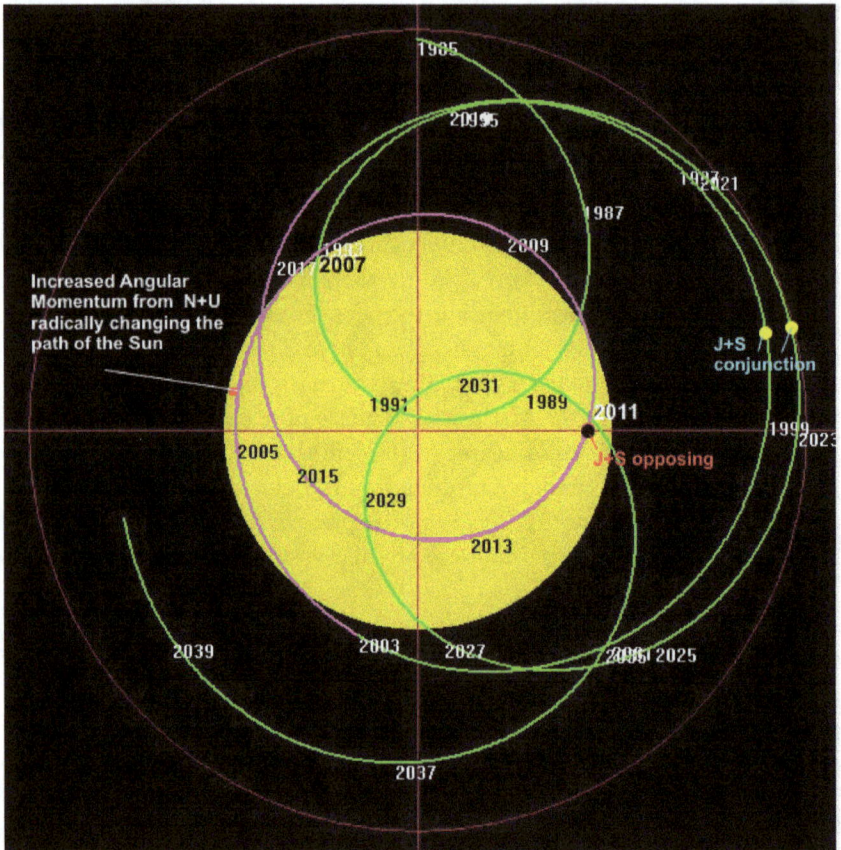

Fig. 3.7   Motion of barycenter of the solar system relative to the Sun. Credit: Wikimedia.

These examples show: the precision needed to find exoplanets with astrometry is in the μas-regime.

The astrometric re-detections of exoplanets were all done with the Hubble Space Telescope (HST): Gl 876 b, 55 Cnc and Epsilon Eridani b.

- For example, the first astrometrically determined mass of an exoplanet Gl 876 b yielded the following values [Benedict *et al.*, 2002]:

semi-major axis: 0.25 ± 0.06 mas,
inclination: $i = 84 \pm 6°$,
parallax: 214.6 ± 0.2 mas,
mass of the primary star: $M_* = 0.32 M_\odot$,
mass of the planet Gl 876b $= 1.89 \pm 0.34\, M_{\rm J}$.

- For the planet orbiting the K2V star ϵ Eridani, the following parameters were found [Benedict *et al.*, 2006]:

semi-major axis 1.88 ± 0.2 mas,
inclination $30.1 \pm 3.8°$,
$M_* = 0.83 M_\odot$,
$M_{\text{Comp}} = 1.55 \pm 0.24 M_J$.

Cool nearby M dwarfs will be investigated for planetary companions with precision astrometry by the GAIA mission.[5]

Another technique applied will be phase-referenced interferometric astrometry. Phase-referenced imaging and micro-arcsecond astrometry (PRIMA) will measure the astrometric wobble of a candidate star due to an exoplanet relative to a close-by "calibrator" star located within the instrument's observing field (1 arcmin in the PRIMA case) [Beust *et al.*, 2011]. The PRIMA is installed at ESO VLTI (Paranal Observatory) and is designed to enable simultaneous interferometric observations of two objects each with size of at most 2 arcsec that are separated by up to 1 arcmin. There are two modes: (i) measure the angular separation between two objects (astrometry mode), (ii) produce images of the fainter of the two objects using a phase reference technique (imaging mode). The PRIMA is managed by Exoplanet Search with PRIma (ESPRI).

The GAIA mission will monitor each of its target stars about 70 times over a 5-year period. It will precisely chart their positions, distances, movements, and changes in brightness. In addition to exoplanets, brown dwarfs, distant quasars, and other objects will also be studied. For all objects brighter than magnitude 15 (4,000 times fainter than the naked eye limit), GAIA will measure their positions to an accuracy of 24 micro-arcseconds. This is comparable to measuring the diameter of a human hair at a distance of 1,000 km. This mission was launched in 2013 and, since 2014, it is located at the Lagrangean point L2, 1.5 million km away from Earth. This location permits an uninterrupted observing mode (Fig. 3.8), and it is an unstable equilibrium point in the Sun–Earth system.

In Fig. 3.9 the number of detected exoplanets is shown; the different colors refer to different detection methods as is explained in the figure caption.

---

[5]GAIA (originally an acronym for Global Astrometric Interferometer for Astrophysics) is a European Space Agency (ESA) space mission in astrometry launched in 2013.

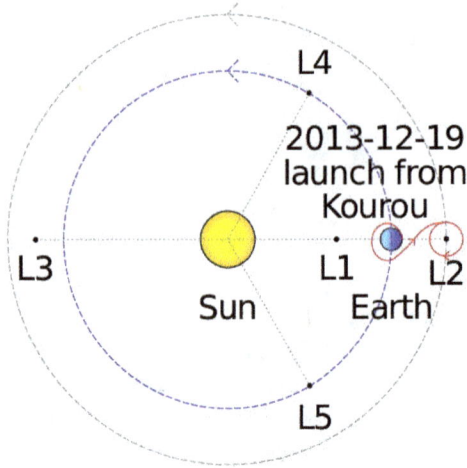

Fig. 3.8   The location of the Lagrangean point L2 where the GAIA satellite observes continuously the sky.

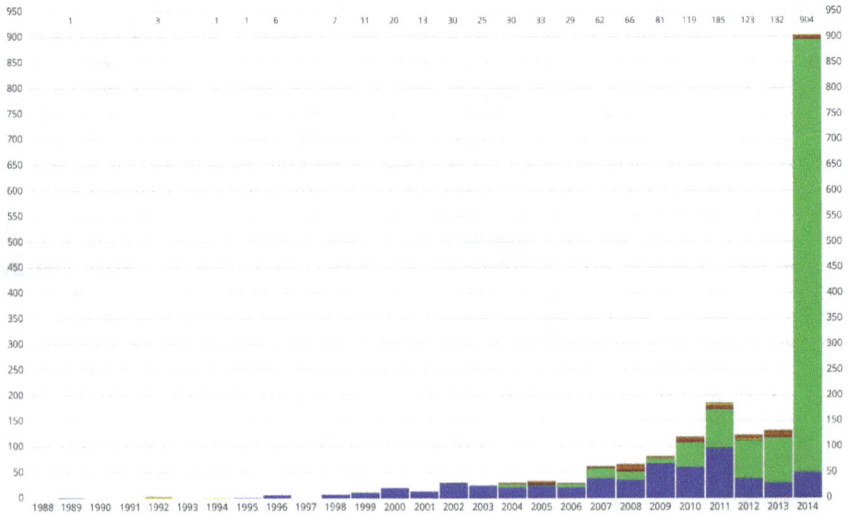

Fig. 3.9   Bar chart of exoplanet discoveries by year, through January 1, 2015, indicating the discovery method using distinct colors: radial velocity (dark blue), transit (dark green), timing (dark yellow), direct imaging (dark red), microlensing (dark orange). Exoplanet data is from the Open Exoplanet Catalogue. Aldaron.

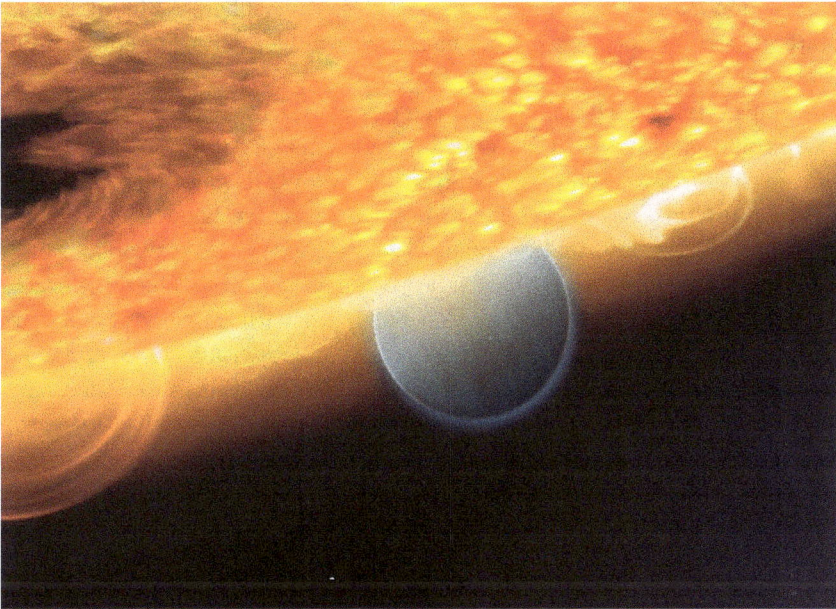

Fig. 3.10   Artist's view of exoplanet orbiting the star HD 189733. Credit: ESA, NASA, M. Kornmesser (ESA/Hubble), and STScI.

3.2   Classification of exoplanets

3.2.1   *Hot Jupiters*

These are objects that have at least the size and mass of Jupiter:

$$M_J = 1.89 \times 10^{27}\,\text{kg}, \qquad R_J = 142{,}984\text{–}133{,}708\,\text{km}.$$

The main difference to the largest planet in our Solar System is however, that hot Jupiters are found extremely close to their host star. If there would be a hot Jupiter in the solar system, its orbit would be inside the orbit of Mercury. Further properties of hot Jupiters are

- they are gaseous planets,
- they have high atmospheric temperatures (because of close distance to host star),
- they are tidally locked, which means their rotation rate equals their orbital period,
- they show unusual hemispheric insolation; there exist permanent day- and night-sides.

Fast winds (sonic speeds) may occur in their atmospheres. The dynamical state of their atmosphere can be measured by the analysis of spectral line profiles.

An example of a hot Jupiter is the object HD 189733b. This exoplanet is a typical hot Jupiter with a semi-major axis of 0.0312 ($\pm$0.0004) AU. Since it is so close to its parent star, it takes only 2 days for one revolution.

The Jupiter-sized planet is too hot for life as we know it to develop. In order to get information about its atmosphere, this exoplanet was observed with the Hubble Space Telescope spectrograph. The wavelengths investigated were in the range 580.8–638.0 nm with a resolving power of $R = 5,000$. Absorption from the Na I doublet within the exoplanet's atmosphere at the nine sigma confidence level within a 0.5 nm band (absorption depth 0.09 $\pm$ 0.01%) was used to measure the doublet's spectral absorption profile. The observations indicate the presence of a high-altitude silicate haze [Huitson *et al.*, 2012]. IR spectra were taken with the Spitzer telescope. This is a lightweight reflector of Ritchey–Chretien design, with a mirror measuring 85 centimeters in diameter. Weighting less than 50 kg, it has been designed to operate at an extremely low temperature to reveal the presence of water. Its launch date was August 25, 2003. Also CO_2 was detected in its atmosphere. The principle is illustrated in Fig. 3.11.

3.2.2   *Neptunian-like exoplanets*

A hot Neptune is an extrasolar planet at a close distance to its parent star, equal to the size and mass of Neptune or Uranus. The first hot Neptune discovered (August 2004) was Mu Arae d (or HD 160691d). The discovery was made with the aid of the HARPS spectrograph (Fig. 3.12), at the European Southern Observatory's La Silla Observatory in Chile. The instrument has been built to obtain very high long-term radial velocities at an accuracy on the order of 1 m/s. Besides the hot Neptune, this system also hosts two Jupiters, the companions b and c [Goździewski *et al.*, 2007], with orbital periods of 600 and 2,500 days.

The central star HD 160691 is a G3 IV-V object with a mass of about 1.1 solar mass and a radius of 1.36 solar radii. It has an age of about 6.34 $\pm$ 0.4 Gyr. The mass of HD 160691d is 14 Earth masses. Its revolution period is about 9 days. It is comparable to the object Gliese 436 b. The semi-major axis of its orbit is 0.09 AU, the eccentricity is 0.172, the periastron[6] is at

---

[6]This is the point on the orbit of the planet nearest to its central star.

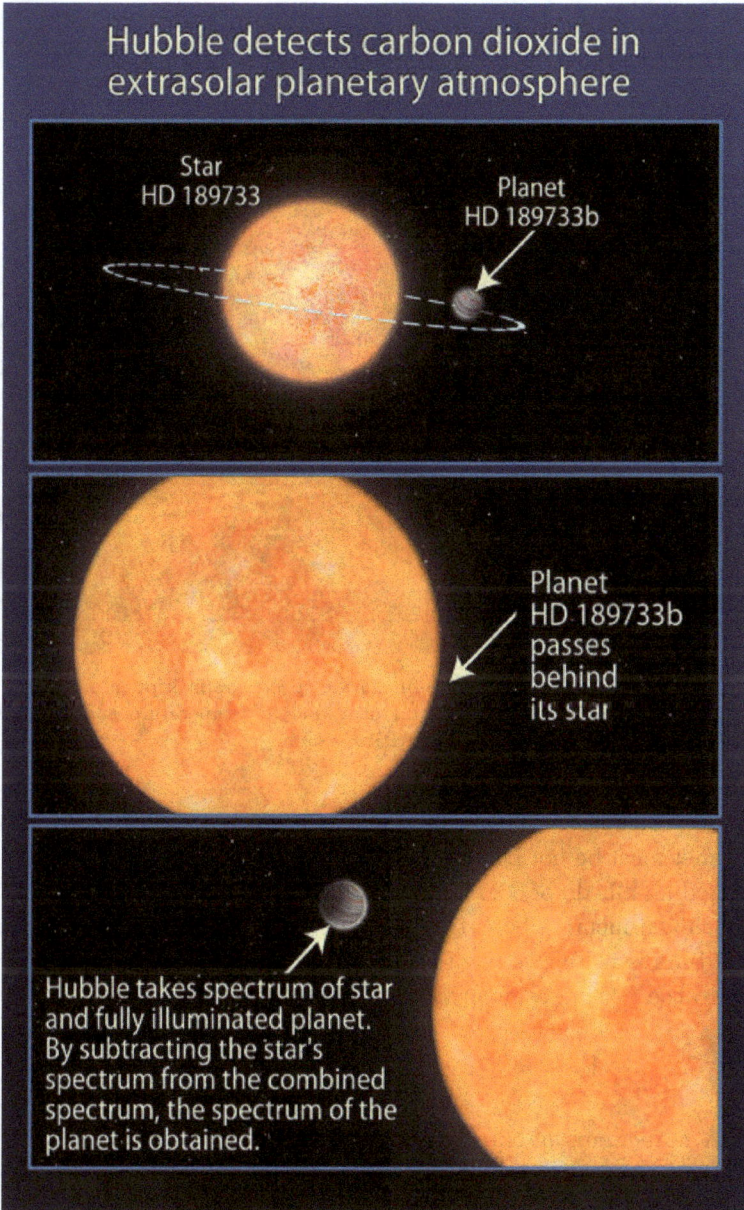

Fig. 3.11   Explanation of how to detect spectral signatures on an exoplanet that transits in front of its host star. Credit: NASA press release.

Fig. 3.12 The HARPS instrument is an echelle spectrograph kept in a vacuum tank (partly removed here) to avoid spectral drift due to temperature and air pressure variations. Credit: ESO, La Silla, the HARPS team.

0.075, and the apastron[7] at 0.106 AU. The orbital period is only 9.63 days. The planet must be hot because of its closeness to Mu Arae. Its discoverers calculated an albedo of 0.35 for it.

We should mention here, that the long-term precision needed to detect such objects is in the range of less than 1 m/s. The contribution of every planet to the reflex motion of the star is given by

$$V_r(t) = K(\cos(\omega + \nu(t)) + e\cos\omega) + V_0, \qquad (3.7)$$

where K is the semi-amplitude, ω the argument of pericenter, $\nu(t)$ the true anomaly, P the orbital period, e the eccentricity and V_0 the offset velocity.

Kepler has found hundreds of Neptune-sized (about 4 $R_\oplus$) planets within 0.5 AU of their stars.

---

[7]Point at which the distance between the planet and its central star reaches maximum.

There could be two ways to explain how these objects formed:

- core-nucleated accretion, and
- outgassing of hydrogen from dissociated ices.

Neptune-size planets at $T_e = 500$ K with masses as small as a few times that of Earth can plausibly be formed by core-nucleated accretion coupled with subsequent inward migration (for more details see Rogers *et al.* [2011]).

3.2.3 *Sub-Neptunes and super-Earths*

This is a class of objects that does not occur in the solar system. The term super-Earth refers only to the mass of the planet, and so does not imply anything about the surface conditions or habitability. The upper mass limit is about 10 Earth masses. In Fig. 3.13 the mass–radius relation for super-Earths is given. Super-Earths have larger masses than the Earth. According to their density there may exist

- Low-density super-Earths; composed mainly of H and He, mini-Neptunes.
- Intermediate-density super-Earths; either have water as major constituent or have a denser core and an extended gaseous envelope. These are called gas dwarfs or sub-Neptune.
- High density: rocky or metallic composition.

In Fig. 3.14 a comparison of sizes of planets with different compositions is given. Earth-like planets have masses of about one Earth mass, mini-Neptunes have masses of about 5 times the mass of Earth.

An example of a super-Earth is 55 Cnc e. It is at a distance of 51 ly, and the planet could have a carbon rich crust (it was even suggested diamond). It is close to its host star, therefore, the surface temperature could be as high as 2,100 K. The bulk density implies a substantial envelope. Such an envelope could either be a low-density, but extended, atmosphere or a compact object, high-density one. In the case of 55 Cnc e, it seems to be the latter case and water plays an important role [Esteves *et al.*, 2017].

Another example is COROT 7b.

3.2.4 *Earth-like exoplanets*

These planets are called Earth-like because they are similar in size, but no Earth-twin has been discovered to date. Due to observational constraints, Earth-like planets can be only detected around nearby stars.

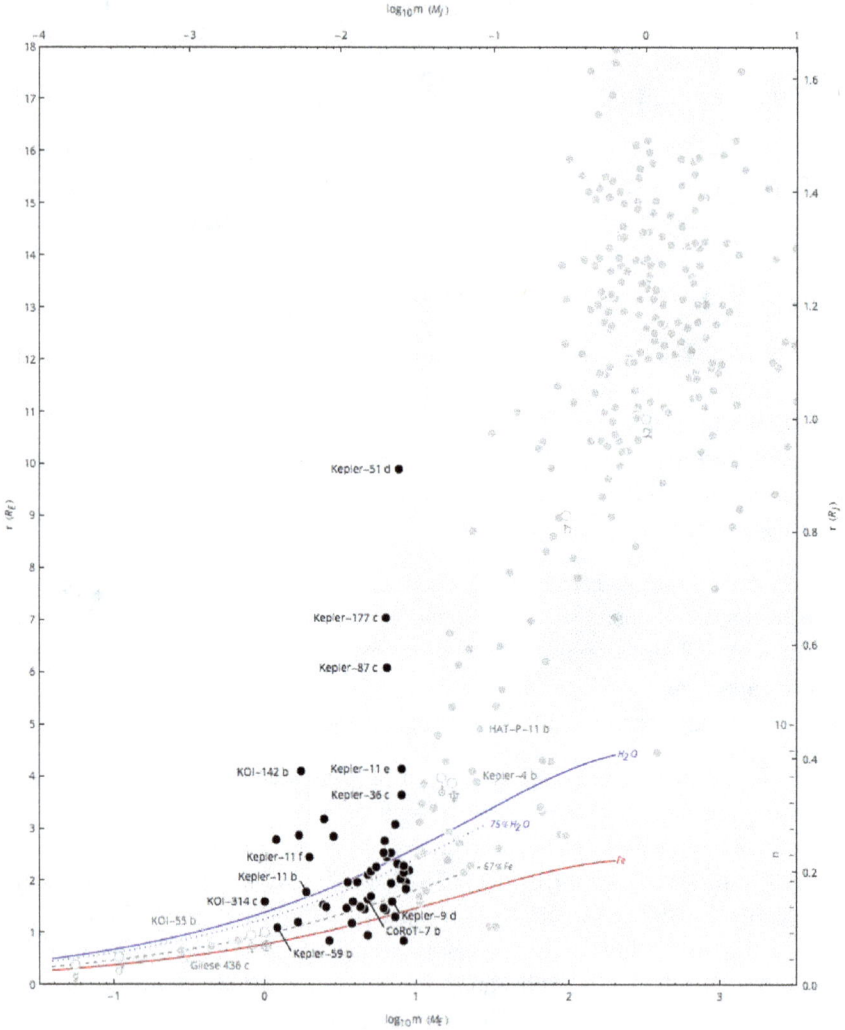

Fig. 3.13   Mass and radius values for transiting super-Earths in context of other detected exoplanets and selected composition models. The "Fe" line defines planets made purely of iron, and "H2O" for those made of water. Those between the two lines, and closer to the Fe line, are most likely solid rocky planets, while those near or above the water line are more likely gas and/or liquid. Planets in the solar system are on the chart, labeled with their astronomical symbols. Credit: Aladaron, Wikimedia Commons.

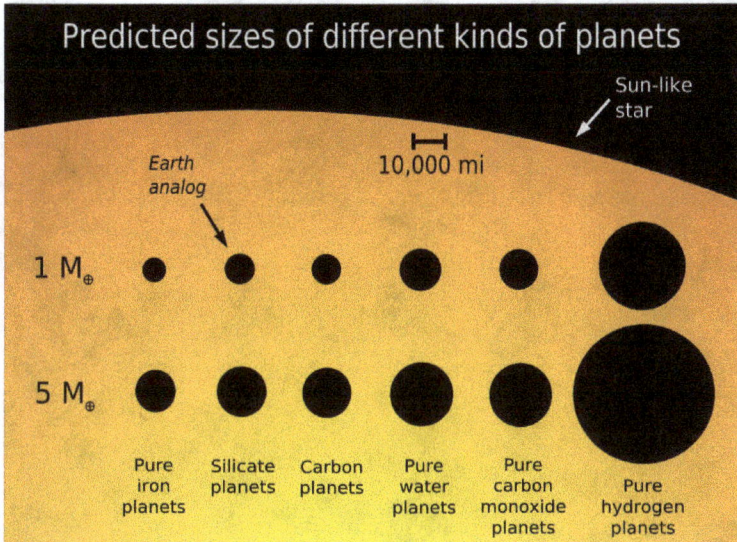

Fig. 3.14 Comparison of sizes of planets with different compositions. Credit: UMD Geology.

In 1957, the German Astronomer W. Gliese published his first catalogue of nearly 1,000 stars located within a distance of 20 pc. Later, several updates appeared. In 1991, the catalogue contained 2803 stars. The most recent update was completed in 2010 [Gliese and Jahreiss, 2015].

We give some examples.

The Gliese 581 system contains five planets. The central star is a red dwarf at a distance of 20.5 light years, to the solar system.

Gliese 581c is one of the most Earth-like planets discovered to date. It is the third of five planets orbiting Gliese 581, completing each orbit in a mere 13 days. Significantly, it is one of the smallest known exoplanets, measuring only 1.5 times the Earth's diameter and only 5 times its mass, and is almost certainly a rock like our own world. It orbits close to the band around its star that is known as the "habitable zone" where water can remain liquid and life as we know it could potentially exist.

The fourth planet in the system, Gliese 581d, is an even better candidate for habitability. In 2009, the discovery of Gliese 581e was announced, with an approximate mass of 1.9 Earth masses.

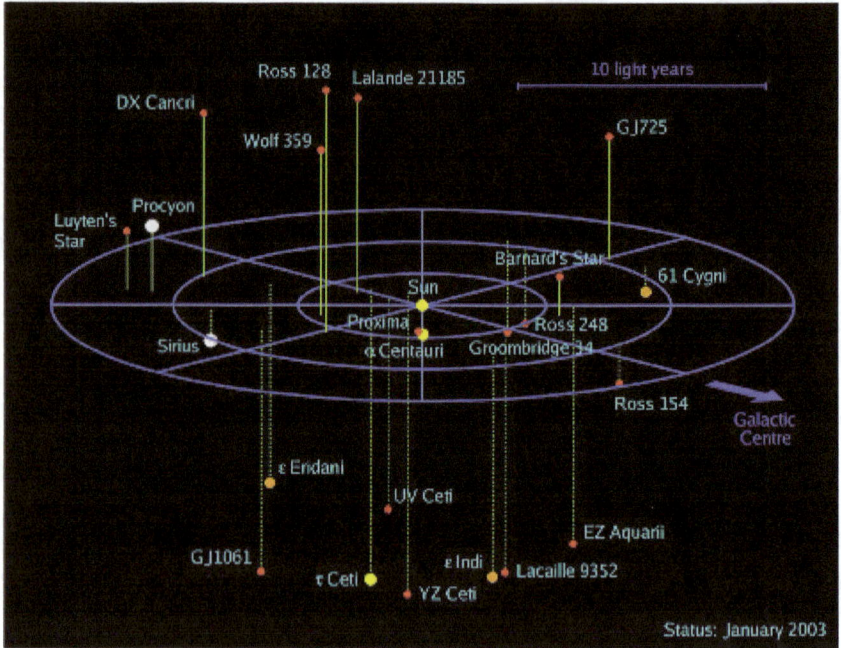

Fig. 3.15   The nearest stars. Credit: ESO.

Planets GJ 581c and GJ 581d are close to, but outside, the conservative habitable zone. Planet c receives 30% more energy from its star than Venus does from the Sun, with an increased radiative forcing caused by the spectral energy distribution of GJ 581. This planet is thus unlikely to host liquid water, although its habitability cannot be positively ruled out by theoretical models due to uncertainties affecting cloud properties and cloud cover. Highly reflective clouds covering at least 75% of the day-side of the planet could indeed prevent the water reservoir from getting entirely vaporized. Irradiation conditions of planet d are comparable to those of early Mars, which is known to have hosted surface liquid water. Thanks to the greenhouse effect of CO_2-ice clouds (also invoked to explain the early Martian climate) planet d might be a better candidate for the first exoplanet known to be potentially habitable [Selsis *et al.*, 2007].

NASA's Kepler Mission uses transit photometry to determine the frequency of Earth-size planets in or near the habitable zone of Sun-like stars. Further examples of exoplanets detected by Kepler are given in Fig. 3.18.

Fig. 3.16   Artist's impression of the Gliese 581 system. Credit: ESO.

An example of a rocky planet is Kepler-10b which has the following parameters: $M_P = 4.56\,M_\oplus$, $R_P = 1.416\,R_\oplus$, and $\rho = 8.8\,g/cm^3$ [Batalha *et al.*, 2011].

The Kepler-20 system (Table 3.2, Fig. 3.17) hosts at least five transiting exoplanets. Kepler-20 e is most likely a rocky planet made of iron and silicates. It is approximately the size of Venus and is the first discovery of a planet smaller than Earth by the Kepler team. It is not habitable because the planet is on an orbit very close to its host star and the equilibrium temperature is as high as 1000 K. Two planets, one Earth-sized ($1.03R_\oplus$) and the other smaller than the Earth ($0.87R_\oplus$), orbiting the star Kepler-20, were reported by Fressin *et al.* [2012]. The gravitational influence of the planets on the parent star is too small to be measured with current instrumentation, and thus transition signals in the lightcurve cannot be confirmed. However, the authors applied a statistical method to show that the likelihood of the planetary interpretation of the transit signals is more than three orders of magnitude larger than that of the alternative hypothesis that the signals result from an eclipsing binary star. Theoretical considerations imply that these planets are rocky, with a composition of

Table 3.2.   The Kepler-20 planetary system.

	Kepler-20b	Kepler-20e	Kepler-20c	Kepler-20f	Kepler-20d
Semi-major axis (AU)	0.04537	0.06335	0.093	0.1378	0.3453
Period (days)	3.69612	6.09849	10.8541	19.5771	77.6118
Mass (M_J)	0.027	0	0.051	0	0.06
Radius (R_J)	0.17	0.0791	0.27	0.093	0.25
Eccentricity	0	0	0	0	0
Inclination (deg)	86.5	87.5	88.39	88.68	89.57

Fig. 3.17   The Kepler-20 system. Credit: NASA/Ames/JPL-Caltech.

iron and silicate. The outer planet could have developed a thick water vapor atmosphere.

3.2.5   *Number of planets in the galaxy*

The number of planets in the galaxy, which is a typical spiral galaxy, can only be estimated by statistics. From such considerations, we can infer, that approximately every star in the galaxy hosts at least one planetary companion. Therefore, the number of planets in a galaxy could become $\sim 10^{11}$ objects. This was investigated in detail in Cassan *et al.* [2012], and we give the most important conclusions from that paper. Most known extrasolar planets have been discovered using the radial velocity or transit

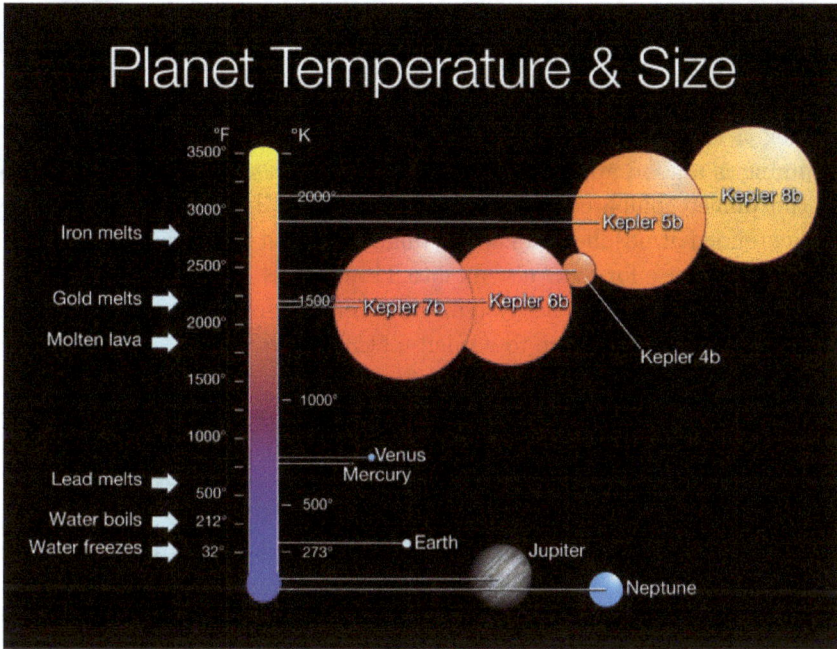

Fig. 3.18   Examples of the first exoplanets detected by Kepler. Credit: JPL.

methods. Both are biased toward planets that are relatively close to their parent stars, and studies find that around 17–30% of solar-like stars host a planet. A population of planets that are unbound or very far from their stars was discovered by microlensing. Within 0.5–10 AU on average,

- $17^{+6}_{-9}\%$ of stars host a Jupiter (0.3–10 M_J),
- and $52^{+22}_{-29}\%$ of stars host Neptune-like planets,
- on average every star has $1.6^{+0.72}_{-0.89}$ planets.

During the past years, planets have been found around every type of star from A to M, including pulsars and binaries.

The false positive rate of Kepler and the occurrence of planets was studied by Fressin *et al.* [2013]. These authors also conclude that 16.5±3.6% of main-sequence FGK stars have at least one planet between 0.8 and 1.25 Earth radii with orbital periods up to 85 days.

Being the leftovers of the stellar formation process, planets appear to be rather ubiquitous, and in reality, the presence of a host star is not always a mandatory circumstance.

3.3 Physics of exoplanets

3.3.1 *Atmospheres: Transmission spectroscopy*

During a transit, the planet passes in front of the star. The amount of dimming is a good approximation for a direct measure of the relative size of the planet compared to the size of the star. If a planet has a dense and extended atmosphere, its actual diameter could be overestimated. This was, e.g. the case for Saturn's satellite Titan, which was thought to be the largest satellite in the solar system; however, it turned out that its actual size was overestimated due to its dense atmosphere.

Extracting the spectroscopic signatures of exoplanetary atmospheres is a challenging task, because they are typically many orders of magnitude fainter than the light from their host stars. Hot Jupiters are particularly accessible to atmospheric characterization via transits and eclipses.

Let us consider a planet with an atmosphere. The stellar light passing through its atmosphere is absorbed by atoms and molecules in the planet's atmosphere. Moreover, the effective size of the planet will appear larger at particular wavelengths of high absorption. A fractional increase in transit depth in an absorption line occurs,

$$\frac{\Delta F}{F} = 2\frac{\Delta R_p}{R_*}\frac{R_p}{R_*}, \tag{3.8}$$

where R_p is the planetary radius, R_* the stellar radius, ΔR_P is the change in radius due to absorption by a molecule. The typical size variation as a function of wavelength is proportional to the atmospheric scale-height

$$H = \frac{kT}{\mu g}, \tag{3.9}$$

where g is the planet's surface gravity, T the atmospheric temperature, μ the mean molecular weight of the gas. For the Earth, this scale-height is about 10 km, for a typical hot Jupiter it is a few hundred km. This corresponds to an increase in transit depth of

- $\sim 10^{-7}$ for Earth,
- $\sim 10^{-4}$ for a typical hot Jupiter.

The propagation of radiation through an atmosphere is described by the radiative transfer equation

$$\mu\frac{\partial I_\lambda}{\partial \tau_\lambda} = I_\lambda - S_\lambda. \tag{3.10}$$

I_λ is the monochromatic intensity which depends on the wavelength, $\mu = \cos\theta$ is the cosine of the incident angle θ relative to the normal, τ_λ is the measured optical depth measured from the top of the atmosphere downward and S_λ is the source function. The source function accounts for both radiation scattered into the line of sight and the thermal emission.

The optical depth is given by

$$\tau_\lambda = \kappa_\lambda ds, \tag{3.11}$$

where ds denotes the geometric depth. Passing a gas element of length ds, the change of the intensity can be simply estimated from the equation of radiative transfer when assuming that the source function becomes zero:

$$\frac{dI_\lambda}{d\tau_\lambda} = -I_\lambda, \tag{3.12}$$

$$I_\lambda = I_{0,\lambda}\exp(-\tau_\lambda). \tag{3.13}$$

Therefore, if $\tau_\lambda = 1$ the intensity has changed by a factor of $1/e = 1/2.718$. In the so-called two-stream approximation, the radiative transfer equation is simplified by an integration over the incoming $(-\pi/2 \leq \theta \leq 0)$ and outgoing hemispheres $(0 < \theta < \pi/2)$. It is assumed that the ratios of various moments of the intensity are constant.

In Malik *et al.* [2017] an open-source radiative transfer code named HELIOS, which is constructed for studying exoplanetary atmospheres, is presented. In its initial version, the model atmospheres of HELIOS are one-dimensional and plane-parallel, and the equation of radiative transfer is solved in the two-stream approximation with anisotropic scattering. The code is publicly available as part of the Exoclimes Simulation Platform (exoclime.net). Several planetary atmospheres were simulated in this study. We give some examples in Table 3.3. In this model, four main infrared absorbers were considered: H_2O, CO_2, CO and CH.

Table 3.3.   Planetary atmospheres simulated with HELIOS.

	GJ 1214b	HD 189733b	WASP-8b	WASP-12b
Orb. separation (AU)	0.01411	0.03142	0.0801	0.02293
Eff. temperature (K)	775	15,175	1,185	3,241
Planetary radius (R_J)	0.25	1.22	1.04	1.78
Stellar temperature	3,252	5,050	5,600	6,300

Several elements and molecules have been identified so far in planetary atmospheres: Sodium, potassium, hydrogen, carbon, oxygen and even water. Tinetti *et al.* [2007], found that absorption by water vapor is the most likely cause of the wavelength-dependent variations in the effective radius of the planet at the infrared wavelengths 3.6, 5.8 and 8 microns. The larger effective radius observed at visible wavelengths may arise from either stellar variability or the presence of clouds/hazes.

In Gao *et al.* [2017] sulfur hazes that may arise in the atmospheres of some giant exoplanets due to the photolysis of H_2S are investigated from a theoretical point of view. The most comprehensive spectral database for transiting exoplanets is given in the ACCESS survey [Apai *et al.*, 2017].

Future observatories for this quest include the James Webb Space Telescope and the new generation of extremely large telescopes on the ground. On a more distant horizon, NASA's study concepts for the Habitable Exoplanet Imaging Mission (HabEx) and the Large UV/Optical/Infrared Surveyor (LUVOIR) missions could extend the study of exoplanetary atmospheres to true twins of Earth. A critical review on the study of exoplanetary atmospheres was given in Deming and Seager [2017]. The theoretical concepts of exoplanetary atmosphere are reviewed by Heng [2017].

The detection of exoplanetary atmospheres in Earth-like planets has also been attempted. Detecting such atmospheres of low-mass low-temperature exoplanets is a high-priority goal on the path to ultimately detect biosignatures in the atmospheres of habitable exoplanets [Southworth *et al.*, 2016]. High-precision HST observations of several super-Earths with equilibrium temperatures below 1,000 K have, to date, all resulted in featureless transmission spectra, which have been suggested to be a result of high-altitude clouds. The detection of an atmospheric feature in the atmosphere of a 1.6 M_E transiting exoplanet, GJ 1132b, with an equilibrium temperature of 600 K and orbiting a nearby M dwarf was announced by Southworth *et al.* [2016]. Observations of nine transits of the planet were obtained simultaneously in the griz and JHK passbands. An average radius of 1.44 ± 0.21 $R_\oplus$, averaged over all the passbands, was found for the planet. This can be decomposed into

- a "surface radius" at ~1.35 $R_\oplus$ and
- a higher contributions in the Z and K bands. The Z-band radius is four sigma higher than the continuum, suggesting the strong detection of an atmosphere.

3.3.2 *Secondary eclipse*

When the planet passes behind the star, the light from it is blocked by the star, and so the observer sees only the light of the star. Comparing the flux before, during and after the eclipse, one can get information about the additional contribution of a planet to the light of the system host-star and planet.

The planet reflects light that is proportional to the albedo, A, the radius of the planet R_P, and the semi-major axis a. The change in flux (coming from the star and reflected flux of a planet; wavelength dependence is neglected here) therefore becomes:

$$\frac{\Delta F}{F} = A \left(\frac{R_P}{a} \right)^2 . \tag{3.14}$$

Let us consider two examples:

- Earth-sized planet orbiting at 1 AU: $\Delta F / F < 1.8 \times 10^{-9}$;
- Hot Jupiter at 0.03 AU: $\Delta F / F < 4 \times 10^{-4}$.

The other additional source is the thermal emission of a planet. This can be estimated as

$$\frac{\delta F}{F} = \frac{F_p}{F_*} \left(\frac{R_p}{R_*} \right)^2 . \tag{3.15}$$

So the contributions of a planet to the total flux comes from (i) reflected starlight (ii) thermal planetary emission. The thermal planetary emission is observed in the IR.

Results for emission spectra of several transiting hot Jupiters were obtained with the Wide Field Camera 3 on the HST. By observing exoplanets during occultation, constraints can be made on the temperature structure and overall energy budget of the planet's atmosphere e.g. in Haynes *et al.* [2014].

3.3.3 *Phase curve*

Like the Moon and inner planets in the solar system (Venus and Mercury), extrasolar planets also show phases. When we observe a planet's thermal emission, it is noted that the variations in light are caused by the temperature distribution in the planet's atmosphere. If both the day- and night-side have the same temperature, no phase variations are seen; the strongest variations are seen if the day-side re-emits all the absorbed stellar radiation before it can be transported to the planets night-side. Therefore, phase curve measurements of the thermal emission from planets allowing the re-distribution of the absorbed stellar light from the day- to the night-side give us information on the density of an atmosphere.

Optical phase curves have become one of the common probes of exoplanetary atmospheres [Oreshenko *et al.*, 2016].

3.3.4 *The mass–radius relation*

A planet mass of 200 Earth mass ($M_\oplus$) seems to be a limit concerning the internal and surface structure of an exoplanet:

- rocky planets $\sim 1 M_\oplus$,
- planets transform from oceanic to gas giants between 1–$200 M_\oplus$,
- above $200 M_\oplus$ only gas giants exist.

Then the following empirical relation can be given:

$$r \sim m^{0.3}, \qquad m < 1, \tag{3.16}$$

$$r \sim m^{0.5}, \qquad 1 < m < 200, \tag{3.17}$$

$$r \sim m^{-0.086}, \qquad m > 200, \tag{3.18}$$

where m denotes the mass (Earth = 1), r the radius (Earth = 1). This is summarized in Fig. 3.19.

3.3.5 *Exoplanet densities and composition*

The density of an exoplanet is important since it determines whether the exoplanet is a terrestrial-like object with a solid surface or a Jupiter-like gas giant. The average density of an exoplanet is given by

$$\bar{\rho} = \frac{M}{4\pi R^3/3}. \tag{3.19}$$

This parameter plays an important role for the mass–radius relation. The following classifications can be made:

- Class I: Ice/gas; this population corresponds to the Saturn/Uranus/Jupiter planet type. The average densities are in the range of 0.3–2.1 g/cm^3
- Class II: Iron/rock; the average densities for these objects are found in the interval 3.6–13.4 g/cm^3. These objects are very near to Earth's average density of 5.5 g/cm^3. The super-Earth population consist of planets with a composition similar to Earth, but often more massive.
- Class III: Degenerate; the third and smallest component, average densities range from 25 g/cm^3 to 34 g/cm^3.

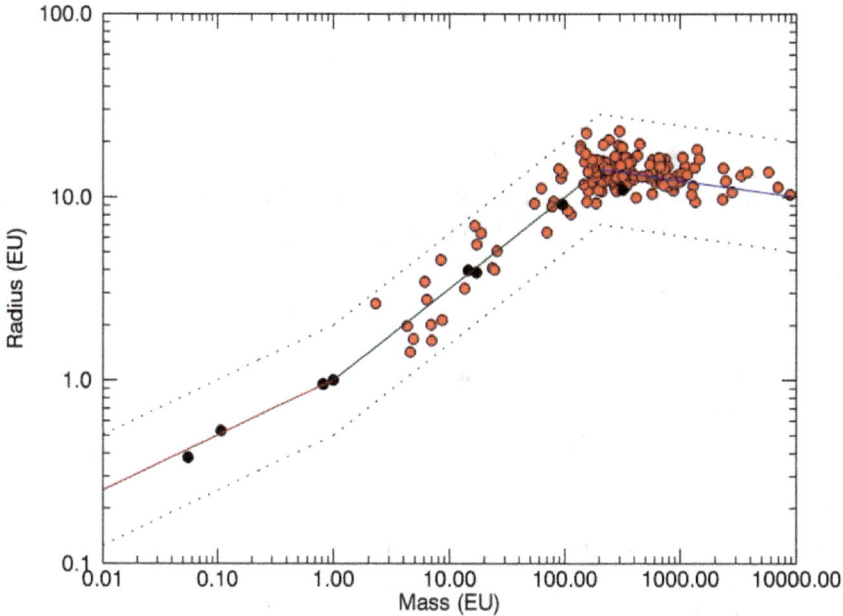

Fig. 3.19 The mass–radius relation for exoplanets. From Planetary Habitability Laboratory @ UPR Arecibo, 2016.

From the sample analyzed, it seems that 80% of all objects are ice giants, 19% belong to class II and only less than 1% to class III [Odrzywolek and Rafelski, 2016].

Planets of size 2–4 $R_\oplus$ have proved to have a wide range of densities, implying a diversity of compositions; however, these measurements did not extend to planets as small as Earth [Howard *et al.*, 2013].

3.4 Exoplanet orbits

The effect of the host star on exoplanets strongly depends on their distance from the host star and on the orbital parameters of the planet. Exoplanetary orbits may not be as stable as in the solar system.

3.4.1 *Stability of planetary orbits*

First we recall some definitions that are needed here in order to understand the principles of the stability analysis. Suppose $\mathbf{f} : \Re^n \to \Re^m$ is a function that takes as input vector $\mathbf{x} \in \Re^n$ and produces as output the vector

$\mathbf{f}(\mathbf{x}) \in \Re^m$. The Jacobian matrix $\mathbf{J}$ of $\mathbf{f}$ is defined by

$$\mathbf{J} = \begin{bmatrix} \frac{\partial f_1}{\partial x_1} & \cdots & \frac{\partial f_1}{\partial x_n} \\ \cdots & \cdots & \cdots \\ \frac{\partial f_m}{\partial x_1} & \cdots & \frac{\partial f_m}{\partial x_n} \end{bmatrix}. \tag{3.20}$$

Let us assume $\mathbf{p}$ is a point in $\Re^n$ and $\mathbf{f}$ is differentiable at $\mathbf{p}$, then its derivative is given by $\mathbf{J_f}(\mathbf{p})$. The best linear approximation of $\mathbf{f}$ near the point $\mathbf{p}$ is

$$\mathbf{f}(\mathbf{x}) = \mathbf{f}(\mathbf{p}) + \mathbf{J_f}(\mathbf{p}) \cdot (\mathbf{x} - \mathbf{p}) + \cdots \tag{3.21}$$

The stability of orbits in such a system as denoted in the above paragraph can be analyzed by using

$$\dot{\eta} = \mathbf{J}(t)\eta \rightarrow \eta = \Delta(t)\eta_0. \tag{3.22}$$

$\mathbf{J}$ is the Jacobian of the right part of the system's equations and $\Delta(t)$ the fundamental matrix of solutions. The deviations $\eta(t)$ remain small, if all eigenvalues lie on the unit circle. Then the spatial periodic orbits are called linearly stable (Floquet's theory).

More details about this analysis can be found in Antoniadou [2016].

In planetary systems, many phenomena can occur such as e.g. a switch from a prograde to retrograde orbit of a planet in a binary system. In Carvalho *et al.* [2016], the secular dynamics of a triple system composed of a Sun-like central star and a Jupiter-like planet, which are under the gravitational influence of another perturbing star (brown dwarf), is studied.

3.4.2 *Tidal locking*

The effect of tidal locking is well known for the Earth–Moon system. The orbital period of the Moon around Earth corresponds to its period of rotation; therefore, we always see the same hemisphere of the Moon from Earth. Most significant moons in the solar system are tidally locked with their primaries, because they orbit very closely and the tidal force exerted by the planet on the satellite increases rapidly (as a cubic function) with decreasing distance.

Many other examples can be found in the solar system. Mercury has a 3:2 spin–orbit resonance, rotating three times for every two revolutions around the Sun. The four Galilean satellites Io, Europa, Ganymede, and Callisto are tidally locked with Jupiter. One hemisphere of Europa is constantly facing Jupiter as a result of tidal locking. Io is in a 2:1 mean motion orbital resonance with Europa and a 4:1 mean motion orbital resonance with Ganymede, completing two orbits around Jupiter for every one orbit completed by Europa, and four orbits for every one completed by Ganymede.

Roughly speaking, the tidal force decreases with r^{-3}, whereas gravitation decreases with r^{-2}. Therefore, tidal locking occurs whenever a planet is close to its parent star.

Proxima b, an Earth-like planet discovered in 2016 that orbits around the star Proxima Centauri is tidally locked, either in synchronized rotation, or with a 3:2 spin–orbit resonance like that of Mercury.

The timescale for a body to become tidally locked can be obtained from the following formula:

$$t_{\text{lock}} \approx \frac{\omega a^6 I \Omega}{3 G m_p^2 k_2 R^5}, \tag{3.23}$$

where ω is the initial spin rate expressed in radians per second, a is the semi-major axis of the planet orbiting a star, $I \approx 0.5 m_s R^2$ is the moment of inertia of the satellite, m_s is the mass of the satellite, R is the mean radius of the satellite, Q is the dissipation function of the satellite, m_p is the mass of the planet, k_2 is the Love number of the satellite.

$$k_2 \approx \frac{1.5}{1 + \frac{19\mu}{2\rho g R}}, \tag{3.24}$$

where ρ is the density of the satellite, $g = GM_s/R^2$ the surface gravity and μ is the rigidity ($3 \times 10^{10}\,\text{Nm}^{-2}$ for rocky objects and $4 \times 10^9\,\text{Nm}^{-2}$ for icy objects).

By assuming a spherical object the above given formula for the tidal locking can be simplified as

$$t_{\text{lock}} \approx 6 \frac{a^6 R \mu}{m_s m_p^2} \times 10^{10}\,\text{yr.} \tag{3.25}$$

Note that

- there is a strong dependence on the semi-major axis ($\sim a^6$!);
- the larger the planet's mass, the shorter the t_{lock} will be.

The maximum orbital distance from main sequence stars where planets can have equilibrium temperatures needed to melt basalts and peridotite depends on the spectral type of the star.

- M0 star or later: $a < 0.01$ AU.
- K0–K5 star: $0.03 < a < 0.04$ AU.

- G0–G5 star: $0.05 < a < 0.07$ AU.
- Sun: $a = 0.055$ AU.

Of course atmospheric circulation of tidally locked exoplanets is different from unlocked planets [Heng *et al.*, 2011].

3.4.3 *Migration of planets*

Many detected exoplanets are so-called hot Jupiters. That means that their mass is comparable to the mass of Jupiter and their orbital period is below 5 days because they are very close to their host stars. The origin of hot Jupiters, large gaseous planets in close orbits around stars, is unknown. Observations suggest that such planets are abundant in stellar clusters; they appear 5–10 times more abundantly there compared to hot Jupiters around stars where no neighboring objects are found. Therefore, a possible explanation is that they are formed due to an instability that was induced by close, passing stars. First, these planets had a high eccentricity, but because of tidal interaction with their host stars, they finally reached a close orbit. Such a scenario was discussed in Triaud [2016].

Planetary migration also occurs when a planet or other stellar satellite interacts with a disk of gas or planetesimals, resulting in the alteration of the satellite's orbital parameters, especially its semi-major axis.

Two mechanisms exist where angular momentum is exchanged.

- Protoplanetary disks are observed around young stars. These disks have typical lifetimes of a few million years. Planets with masses greater than one Earth mass that are formed inside such a gas disk exchange angular momentum. Their orbits will therefore change, in most cases the sense of migration is inward, which means the semi-major axis decreases. However, outward migration is also possible.
- Planetesimal disk interaction: massive protoplanets and planetesimals interact gravitationally. In our solar system, an outward migration of Neptune is believed to be responsible for the resonant capture of Pluto and other Pluto-like objects (Plutinos) into a 3:2 resonance with Neptune.

Migration of close-in planets can also be explained by planet–planet interactions (as discussed in Triaud [2016]).

Chapter 4

Stars and Stellar Activity

In this chapter we will give an overview on basic stellar properties and then discuss stellar activity in general. We will emphasize on cool stars since these stars are the most promising candidates to host planets that are habitable.

4.1 The Sun and stars

4.1.1 Location of the Sun in the galaxy

Typical galaxies contain some 10^{11} stars. Our galaxy belongs to the so-called local group of galaxies. The small and large Magellanic cloud are two small dwarf galaxies which are satellites of our system. The closest large galaxy is the Andromeda galaxy, which is at a distance of more than 2.5 million light years.

Many galaxies appear as spiral galaxies. Young bright stars are found in the spiral arms, while older stars are seen in the center and in the halo of the galaxy. An example is given in Fig. 4.1.

The Sun is located in the Milky Way galaxy at a distance of about 27,000 ly from the galactic center. Our galaxy contains more than 2×10^{11} solar masses (i.e. at least as many stars). The mass of a galaxy can be inferred from the rotation of the stars of the system. At the location of the Sun in the galaxy (Fig. 4.2), one period of revolution around the galactic center is about 200 million years.

4.1.2 Galactic habitable zone (GHZ)

The location of the Sun in the galaxy is also an important factor for habitability on planets. A so-called galactic habitable zone exists. Close to

Fig. 4.1   A typical spiral galaxy. Seen from a distant galaxy, the Sun would be located in our galaxy in one of its spiral arms. Credit: A.H., private observatory.

the galactic center, the density of stars becomes high and the probability of a nearby supernova explosion of a massive star is strongly enhanced. If life would have evolved on some planet around a star close to the galactic center, there is strong risk that it would be destroyed by the intense short-wavelength radiation from such an event.

According to Lineweaver [2007], there are three prerequisites for the evolution of life:

(1) Enough heavy elements to form terrestrial planets,
(2) sufficient time for biological evolution and
(3) an environment free of life-extinguishing supernovae.

Massive stars near the galactic center only have a relatively short lifetime (far below 1 billion years) and, as mentioned above, there is a high risk

Fig. 4.2 The location of Sun in the galaxy as seen from outside. Credit: Caltech.

of a nearby supernova explosion. Another more optimistic view on galactic habitable zones was given by Prantzos [2008].

4.2 Stars: Energy generation and energy transport

The only information we can directly obtain from a star is its radiation and position. In order to understand the physics of stellar structure, stellar birth and evolution, we have to derive quantities such as stellar radii, stellar masses, composition, rotation, magnetic fields, etc. Information on how these parameters can be obtained can be found in any astronomy textbook.

4.2.1 *Timescales*

The gravitational potential can supply the required amount of stellar energy for relatively short timescales during the whole stellar evolution. The release of gravitational energy is governed by the Kelvin–Helmholtz timescale:

$$\tau_{KH} \sim \frac{GM^2}{RL}. \tag{4.1}$$

G is the gravitational constant, M, R, and L are, respectively, the total mass, radius and luminosity of the star. For the Sun $\tau_{KH} = 3.0 \times 10^6$ years.

The nuclear timescale is given by

$$\tau_{nuc} \sim \frac{Mc^2}{L}, \tag{4.2}$$

and for this we get 10^{10} years for the Sun.

Gravitational contraction plays a role for energy generation in stars only during short episodes during their evolution, but it can become important for giant planets like Jupiter.

4.2.2 *Energy transport*

In the interiors of cool stars, energy is transported by radiation, that is, by photons. The mean free path of a photon can be estimated with

$$l_{ph} = \frac{1}{\kappa_{rad}\rho}, \tag{4.3}$$

where κ_{rad} is the mean radiative opacity coefficient due to interactions of photons with particles. For stellar interiors $\kappa_{rad} \approx 1\,cm^2g^{-1}$. For the Sun, $l_{ph} \approx 1\,cm$. This is much smaller than the stellar radius.

4.2.3 *Radiative flux, convection*

The radiative flux is given by

$$F_{rad} = -\frac{4acT^3}{3\kappa_{rad}\rho}\nabla T. \tag{4.4}$$

In the case of spherical symmetry

$$\frac{dT}{dr} = -\frac{3}{4ac}\frac{\kappa_{rad}\rho}{T^3}\frac{L}{4\pi r^2}. \tag{4.5}$$

The condition for the stability against convection is given by the Schwarzschild criterion

$$\left(\frac{d\rho}{dr}\right)_{\text{int}} > \left(\frac{d\rho}{dr}\right)_{\text{ext}} , \tag{4.6}$$

where "int" denotes the change of internal density of a moving mass element while it rises by a distance dr, "ext" indicates the spatial gradient in the star.

The stability of stars is important for their evolution. Cool stars develop a far reaching convective zone and they become very active. Since a habitable planet must be closer to a cool star than to a hot star, it becomes strongly exposed to the activity of its cool host star.

4.3 Stellar Spectra, the Hertzsprung–Russells diagram

4.3.1 *Spectral classes*

According to their spectra, stars can be classified (Table 4.1). O stars are the hottest, M stars the coolest; the number of absorption lines increases from O to M. The fact that more absorption lines are seen in cooler stars is only because of their temperature; the overall chemical composition of the stars is nearly identical.

In Table 4.2 the effective temperature, T_{eff}, is given for the different spectral classes.

4.3.2 *The Hertzsprung–Russell diagram*

In the Hertzsprung–Russell diagram (1913 H. N. Russell added to the work of E. Hertzsprung) (HRD), the temperature of stars is plotted versus brightness. The temperature of a star is related to its color: blue stars are

Table 4.1.   Spectral classification of stars.

O	Ionized He, ionized metals
B	Neutral He, H stronger
A	Balmer lines of H dominate
F	H becomes weaker, neutral and singly ionized metals
G	Singly ionized Ca, H weaker, neutral metals
K	Neutral metals molecular bands appear
M	TiO, neutral metals
R,N	CN, CH, neutral metals
S	Zirconium oxide, neutral metals

Table 4.2.   Effective temperature as a function of spectral type.

Spectral type	O	B0	A0	F0	G0	K0	M0	M5
T_{eff} [K]	50,000	25,000	11,000	7,600	6,000	5,100	3,600	3,000

hotter than red stars. In the HRD the hottest stars are on the left side. The relation between spectral type and stellar temperature is given in Table 4.2. The temperature increases from right to left. Stellar brightness is given in *magnitudes*.

$$\text{magnitude} = \text{const} - 2.5 \log(\text{Intensity}) \tag{4.7}$$

The magnitude scale of stars was chosen such that a difference of 5 magnitudes corresponds to a factor of a 100 in brightness. The Sun has $-26.^{\text{m}}5$. The faintest stars that are visible to the naked eye have a magnitude of $+6.^{\text{m}}0$.

Since apparent magnitudes depend on the intrinsic luminosity and the distance of a star, absolute magnitudes have been defined in order to characterize the true brightness of a star: the absolute magnitude of a star (designated by ^{M}) is the magnitude a star would have at a distance of 10 pc (= 32.6 ly). In the HRD, we can plot stellar absolute magnitudes as ordinates instead of stellar luminosities. The relationship between m and M is given by:

$$m - M = 5 \log r - 5 \tag{4.8}$$

where r is the distance of the object in pc. The Sun has $M = +4.^{\text{M}}5$; seen from a distance of 10 pc it would be among the fainter stars visible with the naked eye.

4.3.3   *Luminosity classes*

The luminosity of a star depends on (a) temperature $\sim T^4$, (b) surface which is $\sim R^2$. At a given temperature, for example, a K star may be a dwarf main sequence star or a giant. Therefore, the spectral class alone does not define the position of a star in the HRD and luminosity classes have been introduced. Class I contains the most luminous supergiants, class II the less luminous supergiants. Class III are the normal giants, class IV the sub giants and class V the main sequence (see Fig. 4.3).

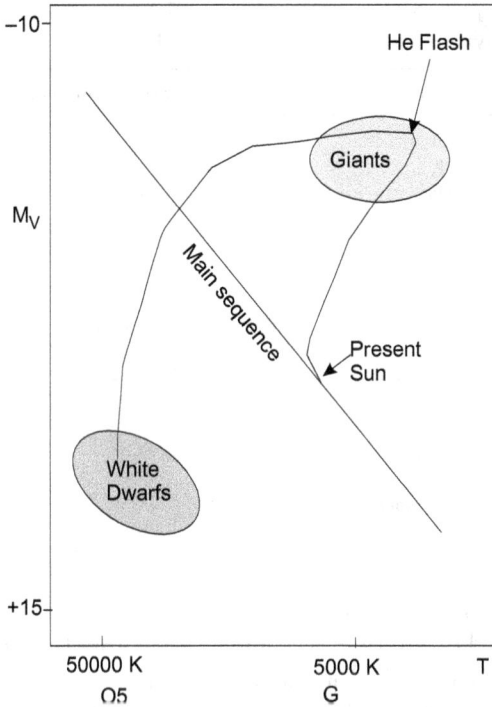

Fig. 4.3 Sketch of the Hertzsprung–Russell diagram with evolutionary path of the Sun.

4.3.4 *Stellar colors*

By putting a prism or a grating inside or in front of a telescope, we obtain the spectrum of a star. Such a spectrum contains many lines, most of which are dark absorption lines. Each chemical element has a characteristic spectrum.

From the spectrum, stellar temperatures can be obtained. The temperature derived from the peak wavelength is called Wien Temperature, the temperature derived from the difference of intensity between two wavelengths (i.e. color) is referred to as color temperature etc. In order to define color, a filter system must be established. The most commonly used system is the UBV system that has three bands that are located in the UV (U), blue (B) and visual (V) to measure the intensity I_ν.

The color of a star is measured by comparing its magnitude through one filter (e.g. red) with its magnitude through another (e.g. blue).

Table 4.3. Central wavelength and bandwidth of the UBVRI filter system.

Name	Meaning	Central λ (nm)	Bandwidth (nm)
U	Ultraviolet	360	66
B	Blue	440	98
V	Visual (green)	550	87
R	Red	700	207
I	Infrared	900	231

Table 4.4. B–V colors and effective temperatures of some stars.

Star	B–V	Effective, T(K)
Sun	+0.6	5800
Vega	0.0	10000
Spica	−0.2	23000
Antares	+1.8	3400

For example, m_V means the magnitude measured with the V filter. Therefore, instead of determining temperatures from the comparison of the spectrum of a star with the Planck law, one can use e.g. color indices. If we calculate B–V, this value will be (see e.g. Table 4.3):

- positive for the cooler star, since they are brighter in V than in B (blue). If the cool star is brighter in V, it means that its magnitude has a lower value and therefore B–V is positive.
- negative for the hotter star. The hotter star is brighter in B than in V; therefore, for the magnitudes in these two bands: $m_B < m_V$ and B–V < 0.

Some examples are given in Table 4.4.

4.4 The evolution of stars

4.4.1 *Distribution of stars in the HRD*

Stars are not randomly distributed in the HRD. This provides a key to understanding the evolution of stars.

- Main sequence stars: most stars are found along a diagonal from the upper left (hot) to the lower right (cool).
- Giants, supergiants: they have the same temperature as the corresponding main sequence stars but are much brighter and must have larger diameters.

• White dwarfs are faint but very hot objects. Therefore, from their location at the lower left in the HRD, it follows that they must be very compact (about 1/100 the size of the Sun).

Why do most of the stars lie on the main sequence? The answer is quite easy: because this denotes the longest phase in stellar evolution. On the main sequence the stars are in hydrostatic equilibrium.

4.4.2 *Age of stellar clusters*

In Fig. 4.4 the HRD of well-known stellar clusters is shown. Stars are formed in clusters. The members of a star cluster are at the same distance to us and have been formed at the same time. These clusters show different turn off points from the main sequence. These turn off points are a proxy for the age of the cluster. More massive stars evolve faster; therefore, we do not expect to find massive stars on the main sequence for older clusters. In this figure we see that e.g. h and χ Persei is a very young cluster. Even massive bright stars are still on the main sequence. On the other hand, M 67 is an

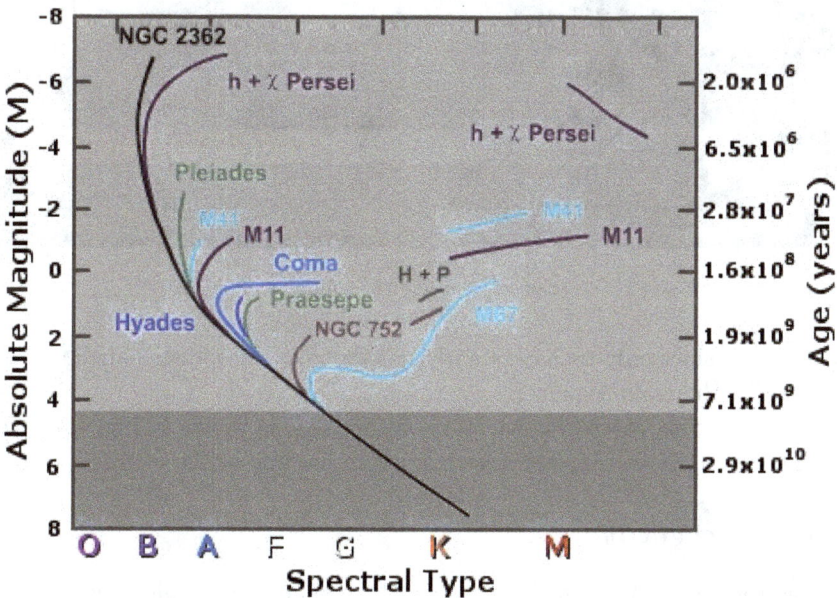

Fig. 4.4 The Hertzsprung–Russel diagram for different stellar clusters. On the right-hand side of the graph the estimated age of the cluster is given derived from the location of the main sequence turn off point. Credit: fr.euhou.net.

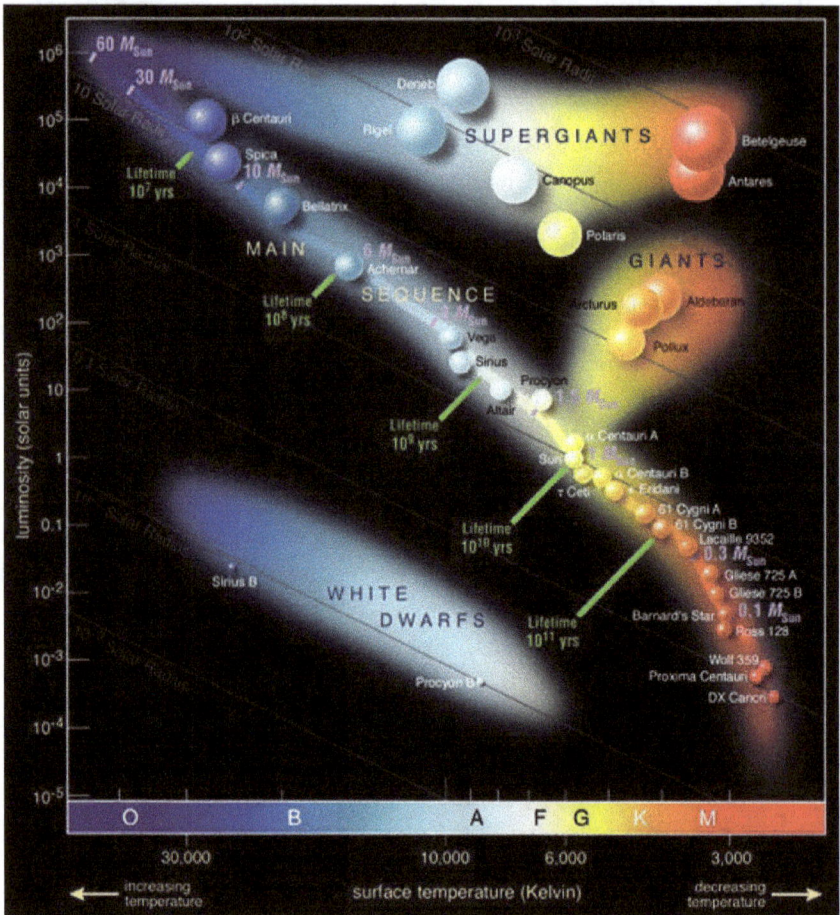

Fig. 4.5   Location of stars and their lifetime on the HRD.

old cluster since only cooler stars with small masses live long enough to still appear at the main sequence.

The location of well-known stars on the HRD is shown in Fig. 4.5. The numbers denote the approximate maximum lifetime of these stars.

4.4.3   *Metallicity*

In astrophysics, all elements heavier than helium are called metals. At the beginning of the universe, just after the Big Bang, there were only the elements hydrogen (called X) and helium (called Y). This primordial helium was produced by thermonuclear fusion during the first 3 min of the evolution of the universe. All elements heavier than helium (called Z) were produced later by fusion reactions in stellar cores. Thus, the stars generate energy by producing metals and the material is dispersed into the universe by stellar winds or explosions. From this material, that is slightly enriched by Z, new generations of stars are formed. Therefore,

- The older generation of stars have lower metallicities (low Z). These are called population II stars.
- The younger generations have higher metallicities. These are called population I stars.

In 1978 a third population of stars was introduced: population III stars which are extremely metal poor stars. These stars are believed to be the first born stars in the universe.

Usually we define

$$X + Y + Z = 1, \tag{4.9}$$

and for example, X is defined as

$$X = \frac{m_H}{M}, \tag{4.10}$$

where M is the total mass of the object (star or nebula). From these definitions we obtain

$$Z = \sum_{i > \text{He}} \frac{m_i}{M} = 1 - X - Y. \tag{4.11}$$

For the Sun we have the values

(1) $X = 0.73$,
(2) $Y = 0.25$,
(3) $Z = 0.02$.

In the paper of von Steiger and Zurbuchen [2016], solar wind compositional data were used to determine the metallicity of the Sun deriving $Z = 0.0196 \pm 0.0014$. Data from photospheric spectroscopy led to lower values, but these values are consistent with results from helioseismology.

Since iron is the easiest element to measure in the visible part of stellar spectra, metallicity is often defined as:

$$[\text{Fe/H}] = \log_{10}\left(\frac{N_\text{Fe}}{N_\text{H}}\right)_* - \log_{10}\left(\frac{N_\text{Fe}}{N_\text{H}}\right)_\odot. \tag{4.12}$$

Often the metallicity is given in dex (decimal exponent). For example, a dex = 1 means that the object has 10 times (10^1) the metallicity of the Sun. A value of -1 means that the objects has only $1/10$ (10^{-1}) the metallicity of the Sun.

For a galaxy, the age–metallicity relation explains a chemical evolution.

4.4.4 *Evolution of the Sun*

The main steps in the evolution of the Sun are (compare with Fig. 4.3):

- Pre-main sequence evolution: before it reaches the main sequence, where it spends most of its lifetime, the Sun forms from a protostellar gas dust cloud. The contracting Sun passes a violent youth, the T Tauri phase.
- At the main sequence, the Sun changes extremely slowly remaining there for about 10^{10} years. In the core H is transformed to He by nuclear fusion.
- The Sun evolves into a red giant, it expands and the Earth becomes part of the solar atmosphere. The expansion starts when all H is transformed to He in the core. Then, a H burning shell supplies the energy. The He flash sets in as soon as He burning begins in the center. The Sun evolves into a red giant for some 10^8 yrs. It extends beyond the Earth's orbit.
- Finally, the Sun becomes a white dwarf which slowly cools.

During its evolution, the Sun dramatically changes its radius:

$1\,R_\odot$ (present Sun) $\rightarrow \sim 10^4\,R_\odot$ (red giant) $\rightarrow 0.01 R_\odot$ (white dwarf).

For space weather, long-term evolutionary effects are negligible. However, it is interesting to investigate them, especially for the early Sun (see Chapter 5 on the faint young Sun problem).

4.4.5 *Solar-like stars*

In order to compare models of stellar and solar evolution with observations, we can study solar-like stars of different ages.

Solar-like stars contain the spectral types late F to late G. The sun-like stars can be subdivided into two groups:

- Solar analogs: Population I dwarfs, properties similar to those of the Sun, can be of different age.
- Solar twins: All physical parameters are almost identical to those of the Sun. The strict criteria to be classified as a solar twin are the following:

  — Temperature within 50 K from that of the Sun (roughly 5,720–5,830 K).

Table 4.5.   Some selected solar twins.

Identifier	Distance	Class	T (K)	Metallicity (dex)	Age (Gyr)
Sun	0.00	G2V	5,778	+0.00	4.6
18 Scorpii	45.1	G2Va	5,790	−0.03	2.9
HD 9986	84	G2V	5,785	+0.09	3.4
HD 150248	88	G2	5,750	−0.04	6.2
HD 164595	91	G2	5,810	−0.04	4.5
HD 195034	92	G5	5,760	−0.04	2.9
HD 117939	98	G3	5,730	−0.10	6.1
HD 138573	101	G5IV-V	5,760	+0.00	5.6
HD 71334	124	G2	5,770	−0.06	5.1
HD 98649	137	G3	5,770	−0.02	4.7
HD 134664	140	G3	5,810	+0.13	2.6
HD 143436	141	G0	5,768	+0.00	3.8
HD 129357	154	G2V	5,749	−0.02	8.2
HD 118598	160	G2	5,800	+0.02	4.3
HD 133600	171	G0	5,808	+0.02	6.3
HD 115382	176	G1	5,780	−0.08	6.1
HIP 11915	190	G5V	5,760	−0.059	4.1
HD 101364	208	G5V	5,795	+0.02	3.5
BD +15 3364	209	G2	5,785	0.07	3.8
HD 197027	250	G3V	5,723	−0.013	8.2

— Metallicity of 89–112% (±0.05 dex) of that of the Sun, meaning the star's proplyd would have had almost exactly the same amount of dust for planetary formation.

— No stellar companion, because the Sun itself is solitary.

— An age within 1 billion years from that of the Sun (roughly 3.5–5.6 Gyr).

The definition of solar twins was given by Cayrel de Strobel *et al.* [1981]. A table of solar twins is given in Table 4.5.

In Table 4.6 we give stars that were selected for the Sun in time project. The star 47 Cas B is the youngest, only 100 million years old, the star 16 Cyg A is the oldest stars, about 8.5 billion years old.

Chemical abundances of solar-like stars can be found in Adibekyan *et al.* [2016].

4.4.6   *Case study: 18 Sco*

One of the best known examples of a solar analog is 18 Sco (HD 1462433, HIP 79672). It was first suggested as a solar twin by Porto de Mello and

Table 4.6. Sun in time stars (Güdel, 2007, *Living Reviews in Solar Physics*, 4).

Identifier	Distance (pc)	Class	Age (Gyr)
47 Cas B	33.5	GV	0.1
EK Dra	33.9	G0 V	0.1
π^1 UMa	14.3	G1 V	0.3
HN Peg	18.4	G0 V	0.3
χ^1 Ori	8.7	G1 V	0.3
BE Cet	20.4	G2 V	0.6
κ^1 Cet	9.2	G5 V	0.75
βCom	9.2	G0 V	1.6
15 Sge	17.7	G5 V	1.9
α Cen A	1.4	G2 V	5–6
βHyi	7.5	G2 IV	6.7
16 Cyg A	21.6	G 1.5 V	8.5

da Silva [1997]. The following parameters were found for this star:

- Surface temperature: 5,433 K,
- Mass: 2.029×10^{30} kg $= 1.02 \, M_\odot$,
- Radius: 703,000 km ($1.03 \, R_\odot$),
- Magnitude: 5.5,
- Distance: 45.3 ly $= 13.0$ pc.

The metallicity of the star is 1.1 times that of the Sun. A metallicity of 1.1 means that the abundance of elements other than hydrogen or helium is 10% greater than that found in the Sun.

In Fig. 4.6 the evolution of the Sun is given from its birth (left-hand side) to its red giant phase (right-hand side). Some solar twins marking different ages are shown in this figure.

The estimation of the given ages of the stars in this table have to be treated with caution. For example, the recent values for 18 Sco were published by Meléndez *et al.* [2014] and indicate that this star is 1.6 Gyr younger than the Sun. Further references and information on this solar-like star can be found in that paper.

4.4.7 *Mass–luminosity relation*

For the main sequence stars, there is a relation between their mass and luminosity:

$$L \sim M^{3.5}. \tag{4.13}$$

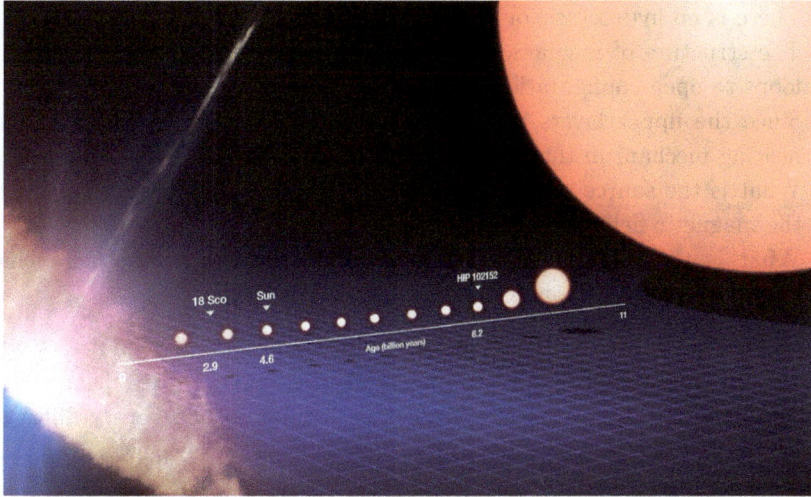

Fig. 4.6 Evolution of the Sun. The position of the solar twins 18 Sco and HIP 102152 = HD 197027 are also marked. The position of 18 Sco corresponds to the state of the Sun's evolution when probably life originated on Earth. Credit: ESO/M. Kornmesser.

From Eq. (4.13) we see that more massive stars are very luminous and therefore they use up their nuclear fuel much more rapidly than low-mass stars like our Sun. Massive main sequence stars that are observed today must have been formed in very recent astronomical history.[1]

The main sequence lifetime of a star can be estimated from the following formula:

$$\tau_{MS} \approx 10^{10} \, \text{years} \left(\frac{M}{M_\odot} \right) \left(\frac{L_\odot}{L} \right) = 10^{10} \, \text{years} \left(\frac{M}{M_\odot} \right)^{-2.5}. \qquad (4.14)$$

4.5 Stellar activity and stellar rotation

4.5.1 *Stellar atmospheres*

The outer atmosphere of stars exhibits a number of interesting phenomena that are not yet fully understood. The physics of these layers gets quite complex because:

- In many cases the simple assumption of spherical symmetry is no longer valid.

---

[1]In some large interstellar nebulae one observes stars that have an age of some 10^5 years.

- There is no hydrostatic or radiative equilibrium.
- The structure of magnetic fields shows a great variety from closed-field loops to open configurations.
- Since the upper layers have higher temperatures, there must be some heating mechanism that is not yet fully understood.
- What is the source of the operating dynamo mechanism that generates the magnetic field?
- What is the relation of this operating mechanism with other stellar parameters such as convection and large-scale motion?

To understand these complex questions for the Sun, it is helpful to study other stars and compare them to the Sun. For example, the extent to which stars in the HRD have outer atmospheres similar to or strongly different from that of the Sun and to study what are the basic parameters that control the stellar activity phenomena are.

The basic parameters could be:

- Mass of a star, M,
- surface gravity, g,
- rotation rate Ω,
- composition, Y, Z.

4.5.2   *Rotation–activity connection*

A pioneering exploration that strongly suggests a relation between stellar activity (measured as a flux in X-ray, L_X) and the rotation was given by Pallavicini *et al.* [1981]. Data from the Einstein satellite for determining the X-ray luminosities were used. L_X denotes the luminosity in X-rays. For stars late-type stars (spectral type F5–M5) and for early-type stars (spectral type <F4), the following relations were found:

$$L_X \sim 10^{27}(v_{\mathrm{eq}} \sin i)^2, \qquad \text{F5–M5}, \qquad (4.15)$$

$$L_X \sim L_{\mathrm{Bol}}, \qquad < \text{F5}. \qquad (4.16)$$

v_{eq} is the measured stellar equatorial velocity, i is the angle of inclination of the rotation axis with respect to the observer. Early-type stars (O3 to A5) have X-ray luminosities independent of rotational velocities, and correlating with bolometric luminosities. Late-type stars of spectral type G to M have luminosities well correlated to equatorial rotational velocities, and are independent of luminosity class.

The stellar activity–rotation period correlation is illustrated in Fig. 4.7.

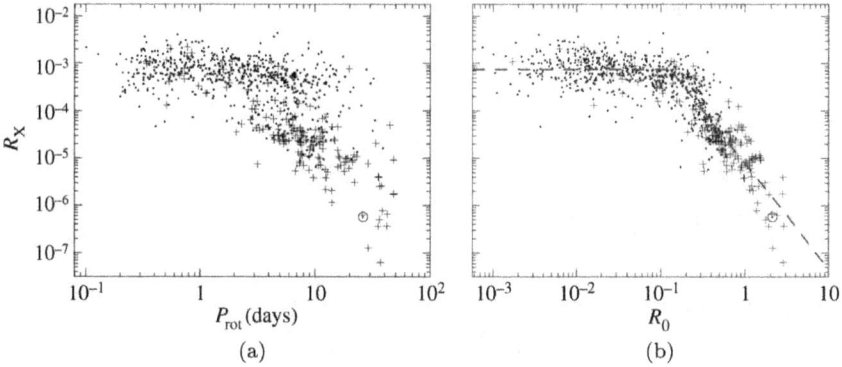

Fig. 4.7   Relationship between L_X/L_{bol} and P_{rot} (a), and L_X/L_{bol} and R_0 (b); from Wright *et al.* [2011]. R_0 is the Rossby number $R_0 = P_{\text{rot}}/\tau$.

An important parameter is the so-called Rossby number R_0. This is defined as the ratio of the convective turnover time to the rotation frequency of the star. Generally, the Rossby number is the ratio of convection to Coriolis forces in hydrodynamics. The convective term can be written as

$$v \cdot \nabla v \sim U^2/L. \tag{4.17}$$

The Coriolis term can be written as

$$\Omega \times v \sim U\Omega. \tag{4.18}$$

These terms appear in the Navier–Stokes equation which, in the incompressible fluid case, becomes

$$\frac{\partial \mathbf{u}}{\partial t} + (\mathbf{u} \cdot \nabla)\mathbf{u} - \nu\nabla^2\mathbf{u} = -\frac{1}{\rho_0}\nabla p + \mathbf{g}. \tag{4.19}$$

$\nu = \mu/\rho_0$ is the kinematic viscosity, μ is the dynamic viscosity, $\mathbf{u}$ is the flow velocity.

As it is seen in Fig. 4.7, there is a saturation at a certain rotation speed. For the unsaturated regime, the authors found the relation

$$L_X/L_{\text{Bol}} \propto R_0^\beta, \qquad \beta = -2.70 \pm 0.13. \tag{4.20}$$

If Ω denotes differential rotation then

$$\Delta\Omega/\Omega \propto \Omega^{0.7}. \tag{4.21}$$

This can be interpreted as a decline of differential rotation of solar-type stars as they spin down.

The two slopes of the stellar rotation-$L_X/L_{\rm Bol}$ relation can be interpreted by different dynamo mechanisms acting.

In the paper of Noyes *et al.* [1984], it was shown that the chromospheric emission of the Ca II H- and K-line were, to a good approximation, a function of the Rossby number for late-type main sequence stars.

4.5.3 *Dynamo number*

Magnetic fields define stellar activity. The generation of magnetic fields on stars can be understood with dynamo theory. The dynamo generation of magnetic fields is parameterized by the dynamo number N_D . The dynamo number is the ratio between magnetic field generation and diffusion terms in the convection zone.

According to Parker [1979], the dynamo number can be written as

$$N_D = \alpha\Omega'd^4/\eta^2. \tag{4.22}$$

In this equation α is the product of the mean helicity of convection and the characteristic turnover time τ_c.

$$\alpha = \langle \mathbf{v} \cdot (\nabla \times \mathbf{v}) \rangle \tau_c. \tag{4.23}$$

The depth gradient of the angular rotation Ω is given by Ω'. η is the turbulent magnetic diffusivity that scales like d^2/τ_C. Since α scales like Ωd we can make the assumption that Ω' scales like ω/d. And

$$N_D \approx (\Omega\tau_c)^2 = R_0^{-2}. \tag{4.24}$$

Therefore, we have found the relation between the Rossby number and the dynamo number

$$N_D \approx R_0^{-2}. \tag{4.25}$$

From the above relation we see that magnetic activity should increase with decreasing Rossby number. Therefore, an empirical relation exists between chromospheric emission and this parameter.

The Rossby number is $R_o = P_{\rm obs}/\tau_c$ (observed rotation period over convective turnover time).

Several empirical formulae were found, for example,

$$\log(P/\tau)f(\langle R'_{\rm HK}\rangle) = 0.324 - 0.400y - 0.283y^2 - 1.325y^3, \tag{4.26}$$

$$y = \log(10^5 R'_{\rm HK}), \tag{4.27}$$

$$x = 1 - (B - V), \tag{4.28}$$

$$\log \tau_c = 1.362 - 0.166x + 0.025x^2 - 5.323x^3, \qquad x > 0,$$
$$(4.29)$$
$$\log \tau_c = 1.362 - 0.14x, \qquad x < 0, \qquad (4.30)$$

where R'_{HK} is the true chromospheric flux.

Finally, there seems to be a very direct link between the rotation period and the age of a star which opens a new and relatively straightforward method for determining stellar age [Meibom *et al.*, 2015; Barnes, 2003].

Gyrochronology: the rotation rates of all cool stars decrease substantially with time as the stars steadily lose their angular momenta (due to stellar wind and magnetic braking).

If properly calibrated, rotation therefore can act as a reliable determinant of stellar ages based on the method of gyrochronology. To calibrate gyrochronology, the relationship between rotation period and age must be determined for cool stars of different masses, which is best accomplished with rotation period measurements for stars in clusters with well-known ages.

4.6 Stellar activity detection

In this section we will discuss several methods that enable the detection of stellar activity. Stellar activity research insight into understanding solar activity and comparisons between solar and stellar parameters can be made.

4.6.1 *Lightcurves*

Huge starspots can be detected by a modulation in the lightcurve due to rotation of the stars. Well-known objects are RS Canum Venaticorum and BY Draconis stars. The starspots on these objects are large, covering about 20% of their surface. The spot temperature was determined near to 3,600 K. The spots are located at intermediate latitudes or even near the poles.

It has already been pointed out that a variation of stellar brightness can be also caused by the transit of a planet. The combination of the two effects, spots and transit leads to a lightcurve as shown in Fig. 4.8.

4.6.2 *Doppler imaging*

The Doppler imaging technique was first used to map chemical peculiarities on Ap stars. In 1983, signatures of starspots were found in the line profiles of the active binary star HR 1099 (V 711 Tau). Doppler imaging is based on the fact that there is a one-to-one correspondence between wavelength

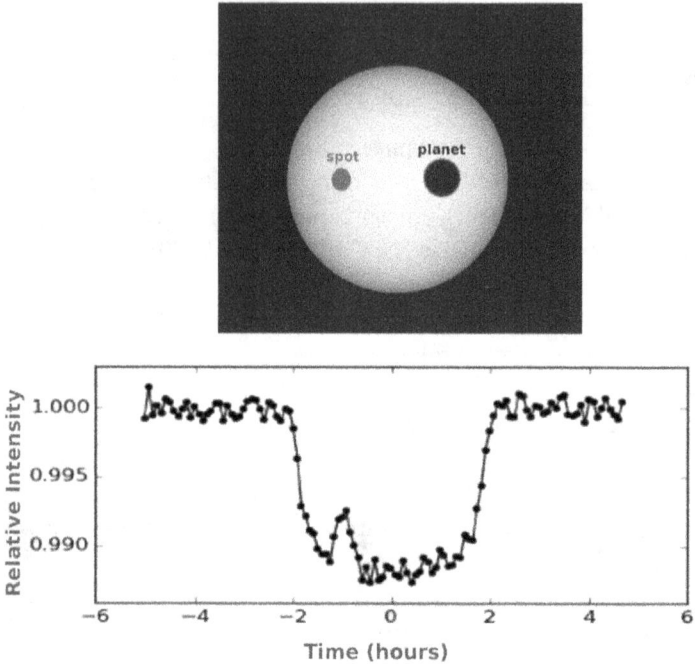

Fig. 4.8   Transit of a planet and starspot.

position across any spectral line and spatial position across the stellar disk. Lines of constant radial velocities are chords across the disk. These are parallel to the star's rotation axis. Any bright or dark region on the surface gives rise to a bump or depression in any spectral line profile at the position corresponding to the radial velocity of the region. The region is carried across the disk by stellar rotation, and therefore the bump or depression also propagates across the line profile.

From this we see immediately: if the rotational Doppler broadening is less or equal to the intrinsic Voigt profile (which determines the line broadening), the mapping effect is not discernible. For stars with $v \sin i$ equal or larger than about $30 \, \mathrm{km \, s^{-1}}$ the Doppler broadening from rotation can be many times greater than the Voigt profile.

The principle of Doppler imaging is shown in Fig. 4.9.

To summarize, the star needs to fulfill several criteria in order to apply the Doppler imaging technique:

- Stellar rotation needs to be the dominating broadening effect, that means $v \sin i$ is sufficiently large. This is a restriction. However, we have to take

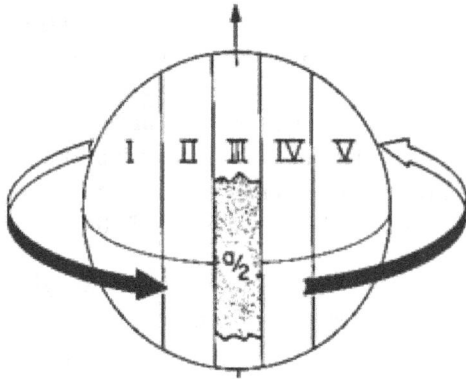

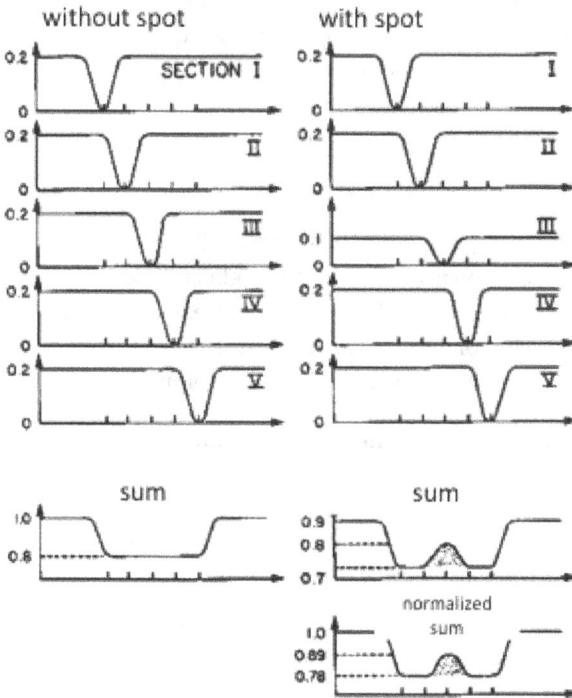

Fig. 4.9   Principle of Doppler imaging. Adapted from Vogt and Penrod [1983].

into account that stellar activity is strongly correlated with their rotation rates. The faster the rotation, the higher the level of activity. Rotation is the key driver for activity. Most active solar-type stars spin fast enough that the rotational Doppler effect becomes the dominant line-broadening mechanism.

- The inclination angle should be between 20 and 70°. This can be easily understood. When $i = 0°$ the star is seen from the pole which, means there is no line-of-sight component of the rotational velocity. The observer sees no Doppler effect. $i = 90°$ means we see equator-on. The Doppler image gets a mirror-image symmetry and it is impossible to distinguish if a spot is on the northern or southern hemisphere.
- The resolution of the spectra should be of the order of 50,000.

Over the last decades, many stars (single, binary dwarf, subgiants and giants) were investigated and spots were detected. By using starspots as tracers, it was even possible to derive the differential rotation of these stars. It was found that most dwarfs and subgiant stars have solar-like differential rotation laws: they rotate faster near the equator than near the poles. The equator overtakes the polar regions on timescales of tens to hundreds of days.

4.6.3 *Zeeman–Doppler imaging*

Magnetic fields polarize the light emitted in spectral lines formed in the stellar atmosphere. This is known as the Zeeman effect [Zeeman, 1896]. The energy states of the electron are split and, therefore, the spectral lines are also split into several components, as is shown in Fig. 4.10.

The splitting depends on whether the observer is looking:

- perpendicular to the magnetic field lines: three components are visible, linearly polarized.

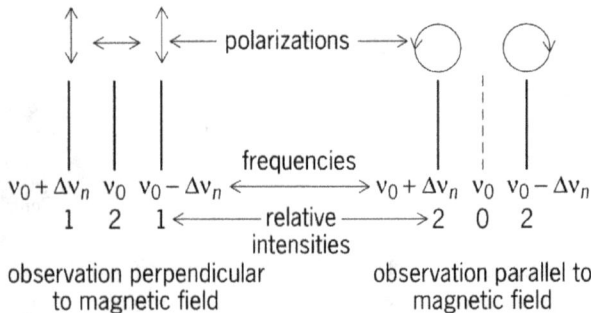

Fig. 4.10 Zeeman effect: when the observers looks parallel to the magnetic field, he sees only two circular polarized components. Credit: McGraw-Hill Concise Encyclopedia of Physics.

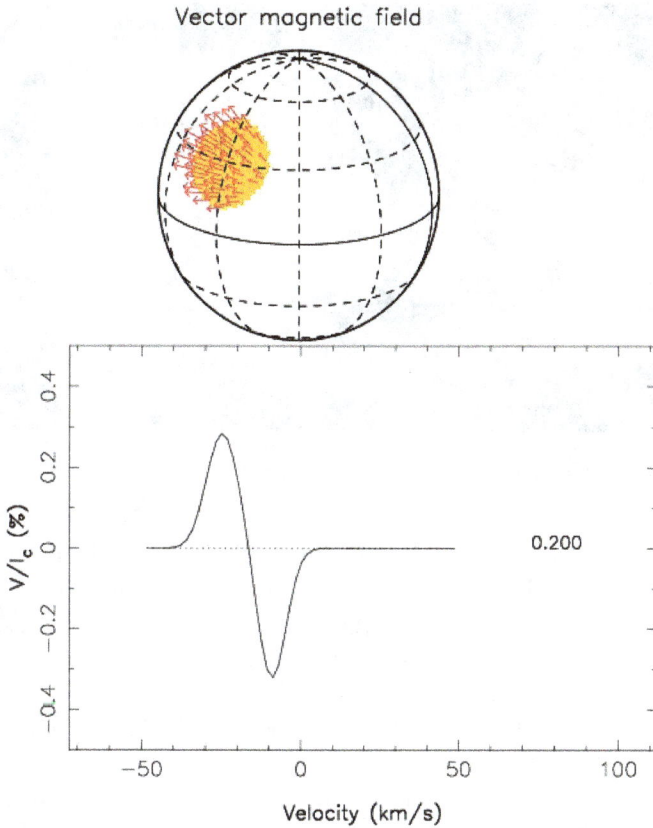

Fig. 4.11   Example of a Stokes-V signal at the blue part because of stellar rotation and a strong magnetic field region.

- parallel to the magnetic field lines: only two components are visible, circularly polarized.

This is shown in Fig. 4.10.

Zeeman–Doppler imaging is based on the combination of stellar rotation and periodic modulation of signatures from the Zeeman effect during such rotations. The method was first applied by Marsh and Horne [1988]. An example of a Stokes-V signal in the line profile is shown in Fig. 4.11. The Stokes-V signal measures the difference between the left- and right-handed polarization signal.

An overview with many references to Doppler and Zeeman imaging is given in Collier [2005]. The problem of reconstructing two-dimensional

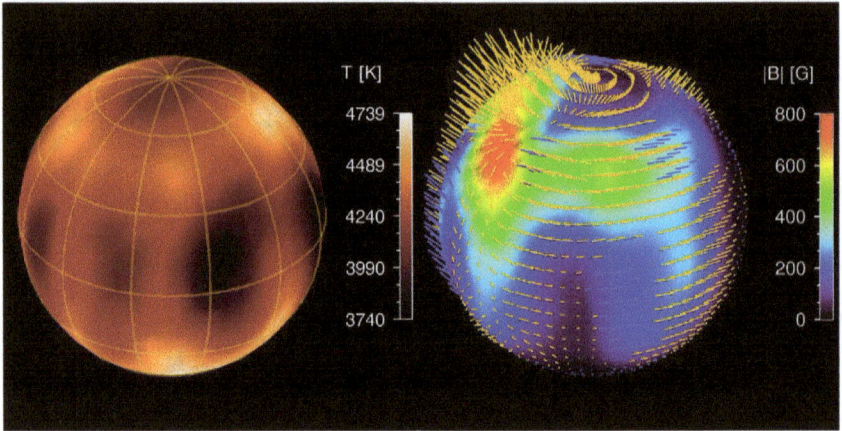

Fig. 4.12　Zeeman–Doppler imaging of II Peg. Credit: AIP.

stellar surface maps from the variability of intensity and/or polarization profiles of spectral lines was reviewed recently by Kochukhov [2016].

Let us consider a Zeeman–Doppler imaging of the star II Peg. This is an active K2 star [Carroll *et al.*, 2009]. The result is shown in Fig. 4.12. In the work of Hackman *et al.* [2016], the three stars AH Lep, HD 29615 and V1358 Ori were analyzed by using the HARPSPol of the 3.6 m ESO telescope. HARPS is the acronym for High Accuracy Radial velocity Planet Searcher. This instrument can measure radial velocities with an accuracy of 1 m/s, the resolving power is 115,000.

Can we expect to find a correlation between the surface brightness maps and the radial magnetic field? In the case of sunspots there is a strong negative correlation: the dimmer the brightness, the stronger the magnetic field. It is difficult to establish such a correlation for stars since many observations have a weak phase coverage.

4.6.4　*Which stars are active?*

As we have seen from the stellar structure description, there are two possibilities for the onset of stellar convection:

- massive hot stars: convective core,
- low massive cool stars: convective envelope. The cooler the star, the larger the extension of the convective envelope.

Surface convection is an important driver for chromospheric activity. In the work of Parker [1970] it is shown that high-temperature chromospheres or coronae can exist only when turbulent (magneto-) convection is present to supply the non-thermal energy needed for heating.

A very general statement can be made: Stellar magnetic activity has been detected in all late-type stars cooler than $T_{\mathrm{eff}} \approx 8{,}300\,\mathrm{K}$. This corresponds to spectral types later than A4V.

Early and middle A stars have very shallow convective layers; they are not expected to produce enough magnetoconvective power to sustain luminous chromospheres or hot coronae. This was investigated by a systematic search for chromospheric emission in the far-ultraviolet (905–1185 Å) spectra of seven main-sequence A stars, based on observations from the Far Ultraviolet Spectroscopic Explorer (FUSE) telescope. This space telescope was launched in 1999 and operated 8 years [Simon *et al.*, 2002]. The transition from radiative to convective envelopes can be determined by searching for high-temperature emissions in the FUSE spectra. The following lines were used:

- from FUSE (1999–2007):
  - — C III at 977 Å and 1175 Å,
  - — O VI at 1032 and 1037 Å.
- from Hubble Space Telescope (1990–), HST (Goddard High-Resolution Spectrograph)
  - — Si III at 1206 Å,
  - — H I Lα at 1215 Å.
- ROSAT, X-ray satellite (1990–1999).

The formation height of these lines requires temperatures between 50,000 and 300,000 K.

Often active stars are grouped as follows:

- Sun-like stars,
- M-dwarfs,
- rapidly rotating stars, i.e. stars in close binaries and young PMS (premain sequence) stars,
- red giants and supergiants.

Considering their optical flux, a variation due to the presence of starspots in their photosphere is found. In the chromospheres of active stars Ca II H and K lines and Balmer emission-line cores are formed. These lines are variable. The presence of a corona is indicated by lines in the EUV–UV and thermal X-ray range and there is also non-thermal radio emission.

4.6.5   *How to measure stellar activity*

First, we stress again that the Sun is the only star whose magnetic activity can be observed in detail. Assuming a strong solar-stellar connection, the activity phenomena that can mainly be observed in late-type stars are decoded using the solar analogous phenomena and their interpretation as reference phenomena.

Stellar activity denotes a group of phenomena observed in the outer atmosphere of late-type stars and there are basically two underlying common phenomena:

- highly complex and structured magnetic fields,
- convective envelope.

Stellar activity phenomena have been first attributed to dissipating acoustic shock waves Schwarzschild [1948]. Later it was recognized that the dissipation of magnetic energy plays the most important role.

Stellar activity can be explained by the transition of complex magnetic fields to a more simple topology.

In many cases, stellar activity is used as a synonym for magnetic activity.

The coupling between differential rotation and turbulent convection produces an intensification of magnetic field and its periodic variation. The dynamo in the Sun is based on a convection zone and a radiative core underneath and is also called $\alpha - \omega$ dynamo. This type of dynamo is also called interface dynamo [Parker, 1977]. A distributed dynamo acts in fully convective stars and brown dwarfs (α^2 dynamo [Chabrier and Küker, 2006]).

TIGRE [Schmitt *et al.*, 2014] is a robotic, automatically operating 1.2 m telescope located in central Mexico, equipped with an Échelle spectrograph with a spectral resolution of 20,000, which covers a spectral range of between 380.0 and 880.0 nm. A main goal consists of monitoring the stellar activity of cool stars, specifically in the emission cores of the Ca II H and K lines. A number of stars with a sampling between 1–3 days over 1 year were monitored for activity [Hempelmann *et al.*, 2016].

The Sun has a steady 11-year cycle in magnetic activity, most well-known by the increase and decrease in the occurrence of dark sunspots on the solar disk in visible bandpasses. The 11-year solar cycle is also manifest in the variations of emission in the Ca II H & K line cores, due to non-thermal (i.e. magnetic) heating in the lower chromosphere. The large variation in Ca II H & K emission allows for studying of the patterns of long-term variability in other stars thanks to synoptic monitoring with the Mount Wilson Observatory HK photometers (1966–2003) and Lowell Observatory Solar-Stellar Spectrograph (1994–present) (see Egeland *et al.* [2016]).

4.6.6 Stellar flares

Stars cooler than our Sun, such as the K and M dwarfs, have flares that seem both surprisingly energetic (flares on active M dwarfs are typically 10–1000 times as energetic as solar flares). The flares are qualitatively different than solar flares, showing strong continuum or "white-light" emission, which resembles a 9,000–10,000 K blackbody superimposed over the quiet spectrum of the star.

With the Kepler Space telescope, a large number of high-energy superflares on cool stars were detected by photometry. These superflares occurred in the energy range of 10^{33}–10^{36} erg. The so-far most energetic flare on the Sun was the Carrington event in 1859; the associated energy was about $\geq 10^{32}$ erg. The proportion of active regions (cool spots) on the surfaces of 279 G-type stars in which more than 1,500 superflares with energies of 10^{33}–10^{36} erg were detected were described in Savanov and Dmitrienko [2015].

White-light emission in solar flares is usually concentrated in small areas near the footpoints of emerging magnetic field lines, spatially associated with hard X-ray emission but does not appreciably affect the integrated optical brightness of the Sun. However, white-light emission on cool stars covers a much larger fraction of the star. The white-light flare emission on the M dwarf YZ CMi was determined to cover 0.22% of the visible disk, the area affected was about 400 mm^2 [Kowalski *et al.*, 2010]. The energy involved was about 6×10^{34} ergs. The 2002 white-light emission on the Sun only covered 20 mm^2.

A search for white-light flares on ~23,000 cool dwarfs in the Kepler Quarter 1 long cadence data was conducted by Walkowicz *et al.* [2011]. In total, 373 flaring stars were identified, some of which flared multiple times during the observation period. M dwarfs tend to flare more frequently but for shorter durations than K dwarfs and that they emit more energy relative to their quiescent luminosity in a given flare than K dwarfs. Stars that are more photometrically variable in quiescence tend to emit relatively more energy during flares, but variability is only weakly correlated with flare frequency.

Parameters of 3,140 flares in 209 stars observed by Kepler in short-cadence mode are given in the catalogue by Balona [2015b]. Flare stars across the H-R diagram were discussed by Balona [2015a]. In that paper, the incidence of flares on cool stars (K–M dwarfs) and on hotter stars (A stars) is compared. The relative number of flare stars does not change very much from cool to hot stars. A cool companion of hot stars would not create flares of the amplitudes observed. Other interesting aspects of flares on stars are:

- Spots on flare stars are generally larger than those on non-flare stars;
- flare stars rotate significantly faster than non-flare stars;
- flare energies are strongly correlated with stellar luminosity and radius.

A homogeneous search for stellar flares has also been performed using every available Kepler lightcurve [Davenport, 2016]. More than 800,000 flare events were detected on more than 4,000 stars. The average flare energy was 10^{35} erg; the net fraction of flares stars increase with decreasing stellar mass.

In the chapters before, we stressed that the rotation rate strongly correlates with magnetic activity and therefore stars with superflares could be assumed as rapid rotators. However, superflares were also found on slowly rotating stars with rotation periods of about 10 days. The Sun is also a slowly rotating star. Could it also happen that the Sun will have some superflare activity? The analysis clearly showed that stars with superflares are heavily spotted, which underlines the magnetic origin of these flares. The maximum energy of superflares is independent of the stellar rotational periods P with the suggestion that the entire range of variations of the flare energies is independent of P.

A survey ox X-ray flares from the 2XMM Catalogue is presented in Pye *et al.* [2015].

4.6.7 *Coronal mass ejections (CMEs)*

Flares and possible CMEs can be observed in the radio wavelengths. The Sun shows type II and type III bursts. These are generated via the plasma emission mechanism. Beams of suprathermal electrons interact with the ambient plasma generating radio emissions at the plasma frequency f_p (the fundamental emission) or at its second harmonic $2f_p$ (the harmonic

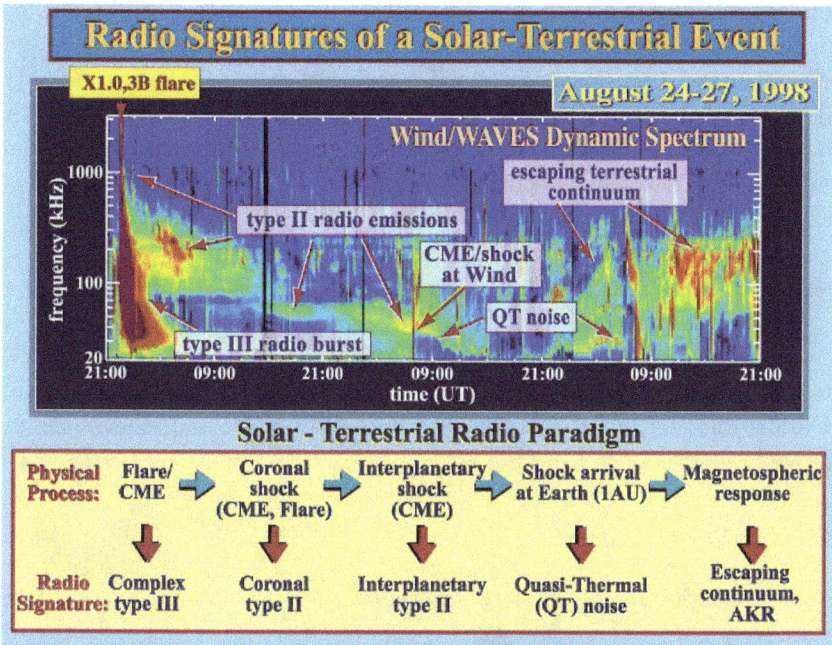

Fig. 4.13 Propagation of a solar CME. The time–frequency diagram of radio emission shows the propagation from higher layers in the solar atmosphere to the interplanetary medium (lower frequencies). Credit: NASA.

emission). The plasma frequency is proportional to the electron density n_e in the plasma:

$$f_p \approx \sqrt{n_e}. \tag{4.31}$$

The plasma frequency drops as the electron density drops, that means, the farther away from the surface of a star (lower density), the lower the plasma frequency will be. In a time–frequency diagram of radio emission we can therefore directly see the propagation of the disturbance (CME) through a star's outer atmosphere (Fig. 4.13).

VLBI stands for very long baseline interferometry. In VLBI a signal from an astronomical radio source, such as a star, is collected at multiple radio telescopes on Earth. The distance between the radio telescopes is then calculated using the time difference between the arrivals of the radio signal at different telescopes. This allows observations of an object that are made simultaneously by many radio telescopes to be combined, emulating a telescope with a size equal to the maximum separation between the telescopes. By such a technique point-like sources like stars can be located and spatially resolved even in the wavelength of radio waves.

The first VLBI images of a main-sequence star, UV Cet A, B both dMe stars, were made by Benz *et al.* [1998]. Stellar flares and coronal mass ejections can be detected by high-cadence spectroscopy of stellar radio bursts. First results from wide-bandwidth VLA observations of nearby active M dwarfs were given by Villadsen *et al.* [2016].

The eclipsing binary Algol showed a continuous X-ray absorption on August 30, 1997. In Moschou *et al.* [2017] it was investigated whether a CME could be the cause for that. A CME mass in the range $2 \times 10^{21} - 2 \times 10^{22}$ g and a kinetic energy in the range 7×10^{35}–3×10^{38} erg was estimated and the results are in reasonable agreement with extrapolated relations between flare X-ray fluence and CME mass and kinetic energy derived for solar CMEs.

4.7 Stellar evolution

Stellar evolution implies that parameters like stellar luminosity or temperature as well as stellar activity change in the long term. These influence on the evolution of planetary atmospheres and may even shift the location of habitable zones in planetary systems.

4.7.1 *Final stages of stellar evolution*

Stellar evolution is mainly determined by just one parameter: the mass of a star. The mass of a star determines its ultimate fate and lifetime on the main sequence in the HRD.

The mass–luminosity relation is slightly different for low-mass and massive stars:

$$L \propto M^{3.5}, \qquad M \leq M_\odot, \tag{4.32}$$

$$L \propto M^{3.0}, \qquad M \geq M_\odot. \tag{4.33}$$

The luminosity is proportional to fuel consumption, while mass reflects fuel supply. The hydrogen burning lifetimes are given by:

$$\tau_{\mathrm{nuc}} = E_{\mathrm{fusion}}/L \propto M/L \propto 1/M^{2.0}. \tag{4.34}$$

Stars spend most of their lifetime burning hydrogen on the main sequence. Core hydrogen burning ceases when 10% of hydrogen has been burnt.

The lifetime of a star can be estimated from:

$$\tau_{MS} \sim 10^{10} \, (M/M_\odot)^{-2.5} \text{ yr}. \tag{4.35}$$

From this we see that:

- The main sequence lifetime of the Sun, $\tau_{MS} = 10^{10}$ yr;
- A star with $M = 10 M_\odot$ has a main sequence lifetime of only 30×10^6 yr;
- A star with $0.5 \, M_\odot$ has a main sequence lifetime of almost 60×10^9 yr.

When core hydrogen burning ceases, stars expand, cool and become red giants. Hydrogen burning continues in a shell outside the inert core. For low-mass stars, the proton–proton chain dominates energy production. The energy production rate is proportional to the temperature:

$$\epsilon_{pp} \sim T^\nu, \qquad \nu = 4. \tag{4.36}$$

For high-mass stars $M > M_\odot$ the CNO-cycle becomes the dominant process and the energy production rate strongly varies with temperature,

$$\epsilon_{CNO} \sim T^{17}. \tag{4.37}$$

Therefore, these stars have a convective core.

As soon as the core is comprised of more than about 10% of He, the He-enriched core contracts and the CNO cycle starts to work even for low-mass stars. The increased thermal pressure (because of the higher temperature) contributes to the expansion of the envelope and the star becomes a red giant. The cooling of the outer, expanding layers causes the outer convection zone to deepen (the star approaches the Hayashi limit where it would become fully convective). The deepening convection zone may carry CNO-cycled material to the surface, this is also called the first dredge-up.

In stars with mass greater than about $0.5 \, M_\odot$, the core becomes hot and dense enough for helium burning to begin ($T \sim 1\text{--}2 \times 10^8$ K):

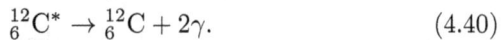

$$^4_2\text{He} + {}^4_2\text{He} \rightarrow {}^8_4\text{Be}, \tag{4.38}$$

$$^4_2\text{He} + {}^8_4\text{Be} \rightarrow {}^{12}_6\text{C}^*, \tag{4.39}$$

$$^{12}_6\text{C}^* \rightarrow {}^{12}_6\text{C} + 2\gamma. \tag{4.40}$$

$^{12}_6\text{C}^*$ indicates a nucleus of carbon-12 in an excited state. In further processes, other elements can form, e.g.

$$^4_2\text{He} + {}^{12}_6\text{C} \rightarrow {}^{16}_8\text{O} + \gamma, \tag{4.41}$$

$$^4_2\text{He} + {}^{16}_8\text{O} \rightarrow {}^{20}_{10}\text{Ne} + \gamma. \tag{4.42}$$

The number of quantum states available to particles is often very large but it is finite. In the very dense interiors of stars, the finite number of quantum states is insufficient for the huge number of particles squeezed into the confined volume. At this stage, matter becomes degenerate and the gas pressure can no longer be described by the ideal gas pressure. Degeneracy sets in as

- the separation of particles becomes less than the de Broglie wavelength for their momentum

$$\lambda_{dB} = \frac{h}{p} \sim \frac{h}{(3mkT)^{1/2}}, \tag{4.43}$$

- the number of particle per unit volume n is higher then the number of available quantum states n_Q.

Electrons become degenerate first, then neutrons also become degenerate at higher pressures (after a supernova explosion).

The equation of state for a classical gas depends on density and temperature:

$$P = nkT. \tag{4.44}$$

For a degenerate gas the pressure becomes:

$$P_{NR} = K_{NR} n^{5/3}, \tag{4.45}$$
$$P_{UR} = K_{UR} n^{4/3}. \tag{4.46}$$

NR stands for non-relativistic, UR for ultra-relativistic.

It is interesting to note that there is a simple mass–radius relation for non-relativistic degenerate stars:

$$R = \frac{4\pi K}{G(4/3\pi)^{5/3} M^{1/3}}. \tag{4.47}$$

The larger the mass, the smaller the radius of a degenerate star.

Late stages of stellar evolution are:

- White dwarfs: for stars with less than about 1.4 solar masses (core mass); this is also called the Chandrasekhar limit. Here, the pressure of degenerate electrons provides equilibrium and no further contraction is possible. White dwarfs have a diameter of about 10,000 km. White dwarfs do not produce energy by nucleosynthesis but they just radiate. This will be the ultimate fate of our Sun and all low-mass stars.

- Neutron stars: when the central mass is above the Chandrasekhar limit, no equilibrium is possible and the star contracts until the pressure of degenerate neutrons, which form from electrons and protons, provides a stable equilibrium.

$$p + e^- \rightarrow n + \nu. \tag{4.48}$$

As soon as the density exceeds $10^{17}\,\mathrm{kg\,m^{-3}}$ the neutrons become a degenerate gas.

Neutron stars are extremely compact objects with diameters of several tens of km.

- Black holes: for stars with central masses of more than 4–5 solar mass, no pressure can prevent from a total collapse and a black hole is formed. The Schwarzschild radius of a black hole is:

$$R_s = \frac{2GM}{c^2}. \tag{4.49}$$

For the Sun, the hypothetical Schwarzschild radius becomes 3 km; for stars with mass M given in solar masses, we have the simple relation:

$$R_s = 3M. \tag{4.50}$$

4.7.2 Violent phases of stellar evolution

Here we briefly discuss violent phases of stellar evolution that strongly influence on planetary systems around them or may even destroy at least close planets.

The transition of low-mass stars to red giants and finally to white dwarfs is rather smooth. At the end of the low-mass star's life, during the red-giant phase, the outer layers of the star are expelled by strong stellar winds. The remaining hot star ionizes the ejected material and we observe a planetary nebula. We know many examples of such objects; for example, the famous Ring Nebula that is at a distance of 2200 ly and has a diameter of 1.3 ly. The central star is a white dwarf with a temperature of about 70,000 K (see Fig. 4.14).

In contrast to the slow expansion and small mass loss of planetary nebulae, there are more violent processes during stellar evolution. Novae are stars that increase in brightness by about 10 magnitudes. The velocity of the expanding material is up to 2,000 km/s. The expanding material amounts to some $10^{-5}\,\mathrm{M_\odot}$. Novae occur in binary systems where one component accretes material from another.

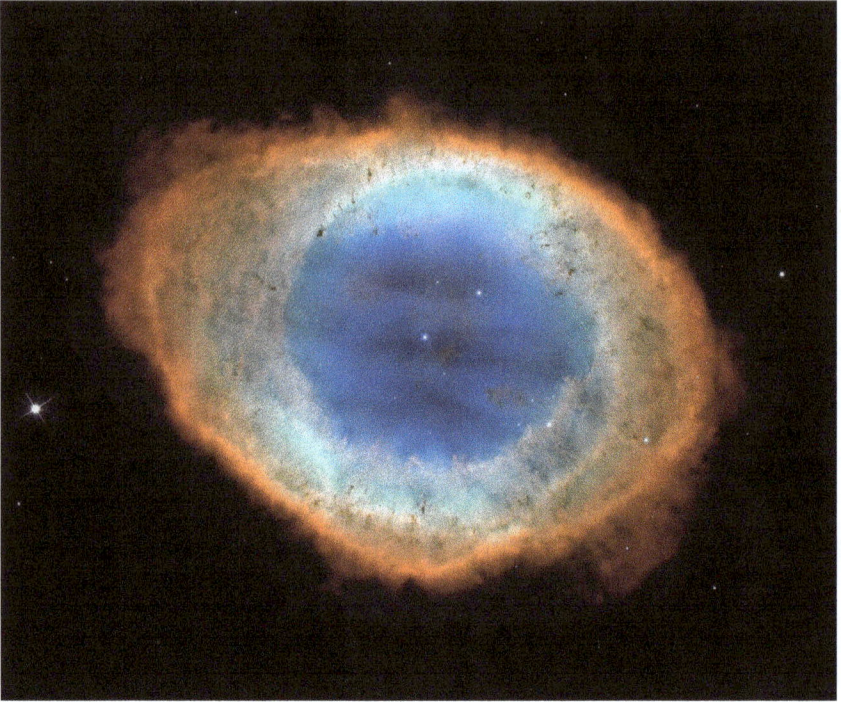

Fig. 4.14   The planetary nebula M 57, also known as Ring Nebula. Credit: NASA/HST.

Massive stars evolve into a supernova by an explosion and finally collapse to a neutron star or a black hole. The velocity of the ejected material is about 10,000 km/s. The energy outburst is about 10^{43} J. Through accretion, low-mass stars can exceed the Chandrasekhar limit; this is called a supernova Type I. During the supernova outburst, the exploding star can become as bright as a whole galaxy, its apparent luminosity increasing by 20 magnitudes. Type II supernovae arise from evolved massive stars (typically O and B stars).

The rate of supernovae in our local Galactic neighborhood within a distance of about 100 parsecs from Earth is estimated to be one every 2–4 million years, based on the total rate in the Milky Way (2.0±0.7 per century).

Observations of open star clusters in the solar neighborhood are used to calculate local supernova (SN) rates for the past 510 Myr. The interstellar matter shows a structuring into bubble and superbubbles. The extension of these bubbles is in the range of 50–150 pc. The lifetime of these structures with lower density is about 10 Myr. The solar system is located in the local superbubble and its structure has been formed by supernova explosions during the last 14 Myr. Therefore, it is reasonable to search for traces and possible effects of such explosions on Earth. Larger dust grains may enter the solar system by penetrating the heliosphere. Typical tracers of such particles are ^{60}Fe and ^{56}Al, which have a half life of about 0.71 Myr. Two enhanced deposition rates of these isotopes have been found in sea sediments:

- An event 8 Myr ago, coincides with ^{3}He and temperature change at the time;
- An event 3 Myr ago occurred at the same time as Earth's temperature started to decrease during the Plio–Pleistocene transition.

More information can be found in Wallner *et al.* [2016].

Peaks in the SN rates match passages of the Sun through periods of locally increased cluster formation which could have been caused by spiral arms of the galaxy. A statistical analysis indicates that the solar system has experienced many large, short-term increases in the flux of Galactic cosmic rays (GCR) from nearby SNe. Could there be a correlation between a high GCR flux and cold conditions on the Earth? By comparing the general geological record of climate over the past 510 Myrs with the fluctuating local SN rates, this question could possibly be answered [Svensmark, 2012]. Surprisingly, a simple combination of tectonics (long-term changes in sea level) and astrophysical activity (SN rates) largely accounts for the observed variations in marine biodiversity over the past 510 Myrs. An inverse correspondence between SN rates and carbon dioxide levels is discussed in terms of a possible drawdown of CO_2 by enhanced bioproductivity in oceans that are better fertilized in cold conditions — a hypothesis that is not contradicted by data on the relative abundance of the heavy isotope of carbon, ^{13}C.

What could be the effect of a nearby supernova explosion? The expected radiation dosage, cosmic ray flux and expanding blast wave collision effects have to be considered. A typical supernova must be closer than $\sim$10 pc before any appreciable and potentially harmful atmosphere/biosphere effects are likely to occur. In contrast, the critical distance for gamma-ray bursts is about 1 kpc [Beech, 2011].

4.7.3   *Stellar evolution and stellar activity*

In this section we shortly review the variation of stellar activity during the evolution of the stars. As we have already discussed, stellar activity can be described by a dynamo mechanism that basically depends on:

- Stellar rotation,
- Stellar convection.

The main sequence lifetime of a star is given by:

$$\tau_{MS} \sim 10^{10} \left(\frac{M_*}{M_\odot}\right) \left(\frac{L_*}{L_\odot}\right)^{-1}. \tag{4.51}$$

Therefore, a solar mass star spends approximately 10 Gyr on the main sequence. We can make a nice comparison here: assume a person lives for 80 years. Then the different phases of stellar evolution correspond to:

- Pre-main sequence evolution (mainly contraction, release of gravitation energy): 30 Myr, 0.2%; this corresponds to about 9 weeks in the lifetime of that person.
- Main sequence phase: this phase is completed at about year 60 of that person's lifetime;
- Post-main sequence time: corresponds to about 20 yrs.

Stars form in groups. Up to an age of about 0.5 Gyr there is a considerable range of rotation rates. Stars spin down via angular momentum loss from magnetized winds. The spin down is greatest for the fastest rotating stars. At an age of about 0.5 Gyr there is a convergence of the rotation rates of a cluster. Beyond that threshold, stars spin down with age according to the Skumanich law:

$$\Omega_* \propto t^{-1/2}. \tag{4.52}$$

This effect also leads to a reduction in the scatter of stellar activity levels with age.

Skumanich [1972] detected this behavior by measuring the Ca II H & K emission. This emission was coupled with a decrease in stellar rotation rate.

Pallavicini *et al.* [1981] reported that the coronal activity of main sequence stars increases with rotation, with X-ray luminosity:

$$L_X \propto (v_* \sin i)^2. \tag{4.53}$$

Assuming a linear dynamo we can also write:

$$L_X \propto B^2. \tag{4.54}$$

The X-ray luminosity scales with magnetic energy density.

The physical reason for such a connection is the coupling of magnetic activity and reconnection with X-ray emission. Nevertheless, X-ray emission does not increase continuously with faster rotation, there is some saturation. The Rossby number R_0 is given by:

$$R_0 = P_{\text{rot}}/\tau_c, \tag{4.55}$$

where τ_c is the convective turnover time. Saturation occurs at

$$L_X/L_* \sim 10^{-3}, \tag{4.56}$$

regardless of spectral type (see e.g. Vilhu [1984]), and for $R_0 \leq 0.1$. Therefore, X-ray emission does not increase continuously with faster rotation; the maximum X-ray emission from main sequence stars saturates at about 1,000 the solar value.

The UV emission also decays over Gyr timescales, although more slowly with age than the higher energy X-ray emission. Concerning FUV wavelengths, Guinan *et al.* [2016] report that the Lyα flux (a good proxy for FUV emission as a whole) decays with stellar age as

$$L_{\text{FUV}} \sim t^{-2/3} \tag{4.57}$$

for early-to-mid spectral type M-dwarfs, more slowly than the $t^{-3/2}$ behavior of the X-ray emission.

The correlation between stellar activity and the rotation rate is illustrated in Figs. 4.15 and 4.16.

Pre-main sequence stars show a different behavior. They do not follow the rotation–activity relation. They show saturated levels of X-ray emission but more scatter in the L_X/L_* compared to the saturated regime of main sequence stars. Pre-main sequence stars are 10^3–10^4 times more X-ray luminous than our present Sun. This high level of X-ray luminosity lasts for the first 100 Myr of their evolution. Then the X-ray luminosity starts to decay. For stars beyond 0.5 Gyrs the decay follows the relation

$$L_X \sim t^{-3/2}. \tag{4.58}$$

The S index is a measure of the strength of the chromospheric emission core of the H and K lines of the Ca II, F_{HK} is its conversion into flux, and R_{HK} is the flux normalized to the bolometric emission of the star.

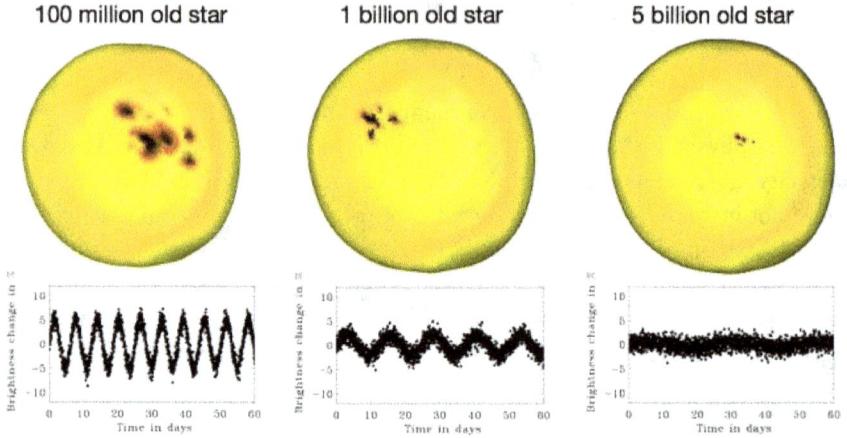

Fig. 4.15 Diagram showing schematically how, as a star ages, its rotation and magnetic activity decrease. Image taken from: http://amazingnotes.com/2011/06/05/how-astronomer-calculate-the-age-of-a-star.

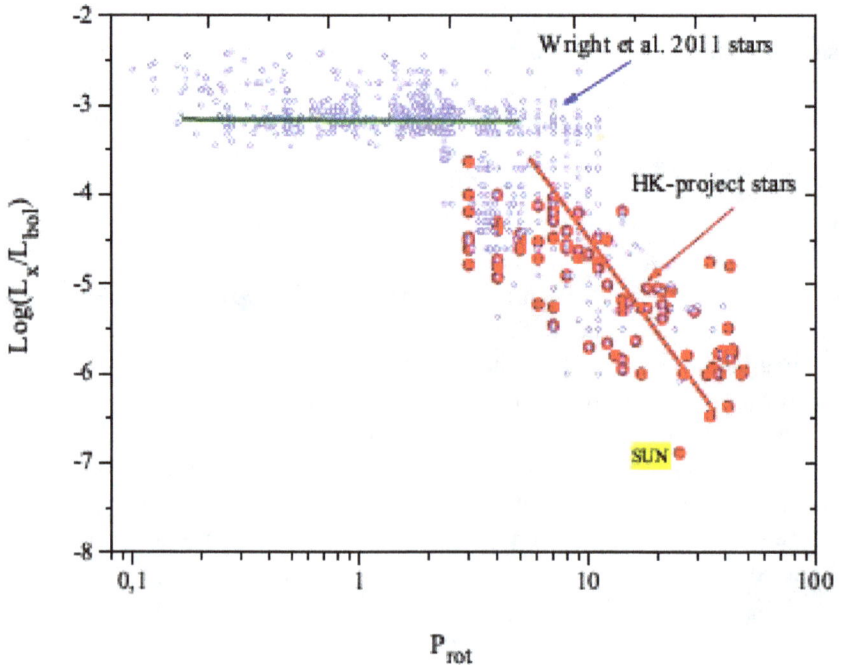

Fig. 4.16 Activity of the chromospheres and coronae of the stars of late (F, G, K) spectral classes, and their cyclic activity is also analyzed on the observational data of solar-type stars from the Mount Wilson HK-project. Credit: E. Bruevich, V. Bruevich and E. Shimanovskaya.

These are the most commonly used proxies for chromospheric activity, and they have been monitored for decades in nearby stars at the Mount Wilson observatory (see also Fig. 4.16).

Chromospheric activity of stars can be measured by the Ca II activity index R_{HK}. This index can also serve as a proxy for the determination of the age of younger stars It seems that this can be used as a proxy for stars younger than 3 Gyr. The Ca II activity and thus the chromospheric activity levels off after about 3 Gyr.

The average longitudinal magnetic field depends on the depth of the convection zone. Stars with a deeper convection zone (like K-stars) have a stronger average longitudinal magnetic field than stars with smaller convective envelopes like F- or G-stars. The longitudinal magnetic field component B_l increases with rotation rate, decreases with age and is well correlated with chromospheric activity [Marsden *et al.*, 2014].

Chapter 5

Habitability

In this chapter we will discuss habitability in detail. First, we give a definition of life. Then the origin of life on Earth is described. Based on life on Earth the concept of habitable zones (HZ) is introduced. Finally, some cases of potentially habitable bodies are discussed.

5.1 Life

The search for life in the universe is based on the assumption that life elsewhere is based on similar conditions that are found on Earth.

5.1.1 *How to define life?*

There are many definitions of life. From the perspective of chemistry, life can be considered as a self-sustaining chemical system that turns environmental resources into its own building blocks. Living beings can be regarded as thermodynamic systems. Like time, entropy runs in one direction. The second law of thermodynamics states that the entropy of a closed system always increases or stays constant. From the point of view of physics, life can be defined as a reduction of entropy. Such entropy reduction occurs locally within an organism and requires external energy or chemical substances. Therefore, on a more global scale, entropy always increases. Such considerations were made by physicists like Schrödinger or Wigner.

For terrestrial life, cells are the building blocks. These cells are mainly composed of the cytoplasm bound by a very thin membrane. Moreover, living cells contain genetic material. This controls their evolution and activities. In many cells, the genetic material, DNA, is found in the nucleus, a spherical structure suspended in the cytoplasm. In simpler forms of life such as bacteria, the DNA is distributed all over the cytoplasm.

The following items summarize some properties of life on Earth:

- Cellular structure of life;
- The need for energy;
- Growth;
- Reproduction;
- Response to stimuli;
- Homeostasis; this Greek word means "staying the same". The internal environment of living things stays relatively constant.
- Metabolism,
- Movement;
- Complexity of organization;
- Adaption to the environment.

From a single cell, life became more complex during its more than 2.5 billion years of evolution on Earth. This is illustrated in Fig. 5.1. On this semilog plot, the complexity of organisms, as measured by the length of functional non-redundant DNA per genome counted by nucleotide base pairs (bp), increases linearly with time.

5.1.2 *Some important inorganic molecules for life*

Water is the most abundant molecule in living organisms, making up between 60 and 70% of the total body weight. Water consists of two

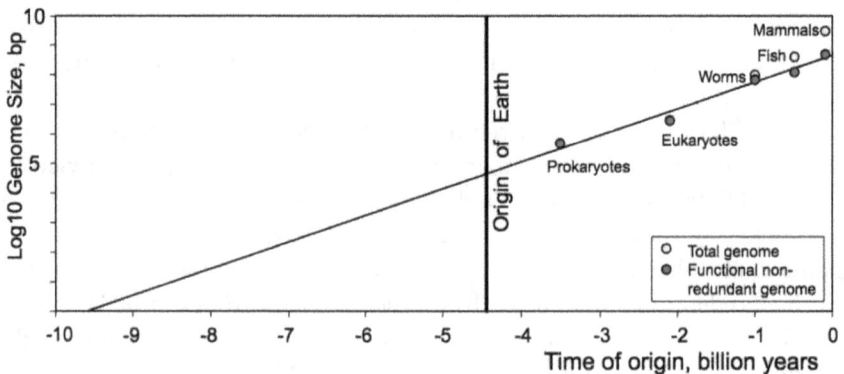

Fig. 5.1   The evolution of complexity of life on Earth. Courtesy: Sharov (2012), arXiv: 1304.3381 [physics.gen-ph].

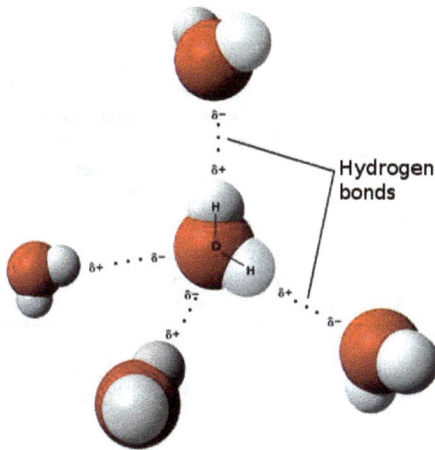

Fig. 5.2   Water molecules consist of oxygen (red) and two hydrogen atoms (small, white). The hydrogen bonds act between the slightly positively charged hydrogen and negatively charged oxygen atoms.

hydrogen (H) and one oxygen (O) atoms. The electrons spend more time circling the larger oxygen atom than the smaller H atom. Therefore, they impart a slight negative charge to the oxygen and a slight positive charge to the hydrogen atoms. The water molecule has a positive and a negative end and becomes a polar molecule. Between the slightly negative oxygen atom of a particular water molecule and the slightly positive charged hydrogen atom of a neighboring molecule, there is a small attraction that is called a hydrogen bond (see Fig. 5.2). These bonds explain the unique properties of water molecules.

Due to their polarity and hydrogen bonding, water molecules are cohesive, enabling them to cling together. This gives water the characteristics beneficial to life. Water is liquid at room temperature and above until 100°C.[1] One further property of water is that the temperature of liquid water rises and falls slowly. Therefore, water protects the organisms from rapid temperature changes and helps maintain the internal temperature of an organism. Water is an excellent temperature buffer. This also helps to moderate the Earth's temperature. Frozen water is less dense than liquid water. As water cools, the molecules come close together, becoming densest

---

[1]This depends on the external pressure; on Mars, temperatures slightly above 0° are reached but, because of the low Martian atmospheric pressure, liquid water would immediately evaporate.

Table 5.1. Significant ions in the body.

Name	Symbol	Significance
Sodium	Na^+	body fluids, muscle contraction, nerve conduction
Chloride	$Cl^.$	body fluids
Potassium	K^+	inside cells; muscle contraction, nerve conduction
Phosphate	PO_4^{3-}	bones, teeth, ATP
Calcium	Ca^{2+}	bones, teeth, nerve conduction, muscle contraction
Bicarbonate	HCO_3^-	acid–base balance
Hydrogen	H^+	acid–base balance
Hydroxide	OH^-	acid–base balance

at $4°C$. Bodies of water like lakes or rivers always freeze from the top to the bottom.

A list of significant ions in the body is given in Table 5.1.

Further important molecules for life on Earth are CO_2, O_2, HCN and others.

5.2 Origin of life

This is still an open question but there are several explanations.

5.2.1 *Panspermia hypothesis*

Did life really originate on Earth or was it transported through space from another object to Earth? Organic compounds are found on meteorites. In Cooper *et al.* [2013] it is demonstrated that organic compounds in a meteorite might even have formed in the early stages of the solar nebula. Such organic material from meteorites has survived the harsh bombardment of cosmic rays in space and the impacts on planetary surfaces.

The panspermia hypothesis suggests that life exists throughout the universe and it is being transported and proliferated to all habitable objects by meteorites and comets. This requires that life survives the extreme conditions in space: (i) extreme cold (down to temperatures near absolute zero), (ii) the hostile radiative environment in the vicinity of stars, (iii) the extreme particle environment (cosmic radiation, high energetic particles), (iv) no nutrients over very long time spans.

It is assumed that bacteria might have traveled long distances in a dormant state and then became active when exposed to habitable conditions on surfaces of planets. Thereafter, evolution started on these planets.

In the 5th Century BC, the Greek philosopher Anaxagoras first mentioned this idea. It was later discussed by Berzelius, Kelvin (1871), Helmholtz (1879) and Svante Arrhenius (1903). Sir Fred Hoyle (1915–2001) and Chandra Wickramasinghe (born 1939) claimed that certain lifeforms may be responsible for epidemic outbreaks, and new diseases when entering the Earth's atmosphere. In that context, the term macroevolution is used.

Panspermia occurs either interstellar (between stars) or interplanetary (between planets). Different transport mechanisms are suggested: Radiation pressure; this was first discussed by Arrhenius. Electromagnetic radiation emitted from stars exerts a pressure upon any surface exposed to it. Light consists of photons which are massless particles. Their energy E is related to their momentum p by the formula:

$$p = \frac{E}{c}. \tag{5.1}$$

Consider a beam of light that is absorbed on a surface. Then, the momentum of the photons of that beam is transferred to the surface, thus transferring momentum to it. Radiation pressure is about 10^{-5} Pa at the Earth's distance from the Sun and decreases by the square of the distance from the Sun. The tiny particles in the dust tail of comets are influenced by the solar radiation pressure and point always away from the Sun.

In the paper of Lingam [2016], two approaches are discussed:

- a self-replication process, endowed with non-local creation and extinction,
- Markov processes and diffusion.

The second process seems to be more robust.

Life that arose via spreading will exhibit more clustering than life that arose spontaneously. A smoking gun signature of panspermia would be the detection of large regions in the Milky Way where life saturates the environment interspersed with voids where life is very uncommon. In a favorable scenario, detection of as few as ~ 25 biologically active exoplanets could yield a 5σ detection of panspermia. Detectability of position–space correlations is possible, unless the timescale for life to become observable once seeded is longer than the timescale for stars to be redistributed in the Milky Way [Lin and Loeb, 2015].

If panspermia is really working then we must take into account that stellar activity can strongly influence on the propagation of life.

5.2.2 *Urey–Miller experiment*

Regardless of whether life has originated on Earth or was brought to Earth from some other place, there must be an explanation for how life could arise from inorganic matter through natural processes. Abiogenesis is the study of these complex reactions and processes.

Modern literature is full of articles on these topics, see e.g. Parker *et al.* [2011].

We start with the Urey–Miller experiment, which was conducted in 1952 and is still considered to be a classic experiment for the origin of life. It was published in 1953 by Stanley Miller and Harold Urey and was conducted at the University of Chicago. In the experiment the following chemicals were used [Miller, 1953]:

- water, H_2O,
- methane, CH_4,
- ammonia, NH_3,
- hydrogen, H_2.

These chemicals were sealed inside a sterile array of glass tubes, connected to a flask of liquid water and another flask containing electrodes. The water was heated and water vapor formed. Lightning was simulated by the electrodes. After some time, the atmosphere was cooled again, and the water could condense. Within a day the mixture turned pink in color; at the end of 1 week, 10–15% of the carbon within the system was in the form of organic compounds and 2% had formed amino acids which make proteins in living cells, with the most abundant being glycine. Sugars and lipids also formed. In all experiments, both left-handed and right-handed optical isomers were created in a racemic mixture. Note that in biological systems most of the components are non-racemic, or homochiral.

About 4 billion years ago, major volcanic eruptions occurred on Earth and they released CO_2, N_2, H_2S, SO_2 into the atmosphere. When these gases were included into the Urey–Miller experiment, more diverse molecules appeared. The early Earth atmosphere could have contained up to 40% of hydrogen and was therefore hydrogen rich. There are several places in the solar system and may be in extrasolar planetary systems, where conditions similar to those of the Urey–Miller experiments are present. However, the original Urey–Miller experiment is increasingly disfavored as an explanation for the origin of Earth life because the primordial atmosphere was probably not that reducing, but still mainly composed of CO_2 and H_2O.

5.2.3 Black smokers

The Earth is geologically active. Tectonic plates move apart, and on ocean basins, hydrothermal vents are found near fissures from which hot water escapes. On land we observe them as hot springs, fumaroles and geysers. Under the sea, these hydrothermal vents may form the so-called black smokers. Black smokers are similar to geysers on the surface but they are found on the floors of the oceans. They appear black due to the condensed minerals from warm or hot mineral-rich water being released into the cold ocean. Black smokers were first discovered in 1977 on the East Pacific Rise using a deep submergence vehicle called Alvin. White smokers are vents that emit lighter-hued minerals, such as those containing barium, calcium, and silicon. These vents also tend to have lower temperature plumes. An example of such a hydrothermal vent is given in Fig. 5.3. On Earth, hydrothermal vents occur in the deep ocean along the mid-ocean ridges: the East Pacific Rise and the mid-Atlantic Ridge (Fig. 5.4). There, tectonic plates move away from each other and new crust is formed. In great

Fig. 5.3 A 5-foot-wide flange, or ledge, on the side of a chimney in the Lost City Field is topped with dendritic carbonate growths that form when mineral-rich vent fluids seep through the flange and come into contact with the cold seawater. Credit: National Science Foundation (University of Washington/Woods Hole Oceanographic Institution).

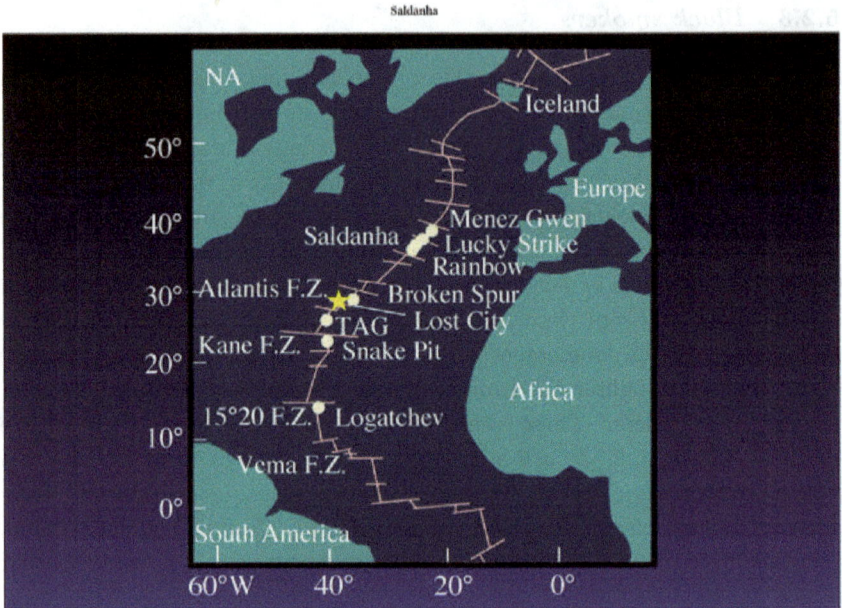

Fig. 5.4   The Mid-Atlantic Ridge. Credit: University of Washington.

ocean depths the ambient water has a temperature of about 2°C. Water that emerges from the vents has a temperature ranging from 60°C to more than 400°C. The pressure is about 218 atmospheres. Pure water can exist in liquid form up to 375°C. At a depth of 3,000 m, the hydrostatic pressure is more than 300 atmospheres because salt water is denser than pure water.

The length of the world ridge-crest system is about 55,000 km. About 10 hydrothermally active sites have been visited.

Near Ascension Island on the mid-Atlantic Ridge, there are three black smokers that have been active since an earthquake occurred there in 2002. The measured temperatures of the water were up to 400°C.

Active hydrothermal vents are believed to exist on Jupiter's moon Europa and other satellites of Jupiter or even Saturn (Enceladus), and ancient hydrothermal vents have been thought to exist on Mars.

Why are these hydrothermal vents important for the origin of life? Hydrothermal vent zones have a density of organisms 10,000–100,000 times greater than the surrounding sea floor. The organisms found there depend on chemosynthetic bacteria for food. Black smokers are the centre of entire ecosystems. There is no sunlight at these depths. Archaea and extremophiles (see Chapter 6) convert the heat, methane and sulfur compounds into energy by chemosynthesis. They are at the base of the food chain. Tube worms form an important part of the community around a hydrothermal vent. They have no mouth or digestive tract, and, like parasitic worms, absorb nutrients produced by the bacteria in their tissues.

Long-term seafloor observatories will allow for the exploration of linkages between volcanism and this newly discovered biosphere in the vicinity of black smokers. Such approaches may provide essential new information about our own planet while providing critically needed insights into how we can explore other planets for life [Kelley *et al.*, 2002].

Wächtershäuser [1989, 1990] proposed the Iron–Sulfur World theory, according to which, life might have originated near hydrothermal vents.

More information about black smokers and chimney chemistry can be found in Von Damm [1990]. The growth of "black smoker" bacteria at temperatures of at least 250°C was investigated in Baross [1983].

5.2.4 *The main stages in the evolution of life on Earth*

In this section we summarize the main stages that seem to be crucial for the evolution of life on Earth. If life appears elsewhere in a similar form, then these stages might be also be similar on such planets.

(1) About 4.5 billion years ago: Formation of the Moon. The Moon was formed by a collision between the early Earth and a protoplanet.
(2) Importance of the Moon for life on Earth: The Moon has a stabilizing effect on the Earth's rotation axis, which stabilizes the climate on Earth. It is also assumed that tides played an important role for the transition of life from sea to land.
(3) About 3.5 billion years ago: Oldest microbial life fossils.
(4) Up to about 2 billion years ago: No free oxygen in Earth's atmosphere.
(5) About 2.7 billion years ago: Sterols = eukaryotes?[2]
(6) About 1.5 billion years ago: Eukaryotes appear.
(7) 750 million years ago: Multicellular eukaryotes.
(8) 600 million years ago: Cambrian explosion.

The cause for the Cambrian explosion of life on Earth is still under debate. By the start of the Cambrian explosion, the large supercontinent

---

[2]Sterols and related compounds play essential roles in the physiology of eukaryotic organisms.

Gondwana, comprising all land on Earth, was breaking up into smaller land masses. This increased the area of the continental shelf, and produced shallow seas, thereby also expanding the diversity of environmental niches in which animals could specialize and speciate. The geology and habitability of terrestrial planets are reviewed in Southam *et al.* [2007].

5.3 Habitable zones

Liquid water is essential to all life on Earth. Therefore, the definition of an HZ is based on the hypothesis that extraterrestrial life would share this requirement. This is a very conservative (but observationally useful and simple) definition.

5.3.1 *Definition*

An HZ can be defined as:

"The orbital region around a star in which an Earth-like planet can possess liquid water on its surface and possibly support life."

However, this is a very crude definition because the planet's surface temperature not only depends on its proximity to its star but also on other factors:

- atmospheric greenhouse gases,
- oceanic circulation,
- internal energy sources (e.g. radioactive decay),
- tidal heating.

These energy sources can maintain a liquid subsurface ocean and a planet could contain life without being within the star's HZ.

HZ can be subdivided into:

- Permanently Habitable Zone (PHZ): the PHZ is the region where a planet always stays within the insolation limits of the corresponding HZ.
- Extended Habitable Zone (EHZ): in contrast to the PHZ, parts of the planetary orbit lie outside the HZ. About 50% of stars are binary or multiple systems. Yet, the binary star–planet configuration is still considered to be habitable when most of the planet's orbit remains inside the HZ boundaries.
- Averaged Habitable Zone (AHZ): this category encompasses all configurations that allow for the planet's time-averaged effective insolation to be within the limits of the HZ.

5.3.2 *Habitability in the solar system*

The extension of a HZ around a star, the so-called circumstellar HZ, depends on stellar parameters such as stellar mass, luminosity, age, evolution as well as on parameters of the object where life is assumed to exist.

What is the extension of an HZ around a star, and how does it depend on the mass of a star and its age?

Let us first consider the radiative flux of a star similar to the Sun. The flux (integrated over all wavelengths) varies with the square of the distance from the star, and therefore:

$$F \sim 1/r^2. \tag{5.2}$$

To illustrate the different fluxes on our neighboring planets Venus and Mars, the following calculation has to be completed: The semi-major axis of Venus is 0.7 AU, the semi-major axis of Mars is 1.4 AU. Then, the solar flux at the orbit of Venus is:

$$F_{\text{Venus}} \sim 1/(0.7)^2 \sim 2. \tag{5.3}$$

On Venus, the solar flux to be expected will be twice that of the flux on Earth; for Mars we obtain:

$$F_{\text{Mars}} \sim 1/(1.4)^2 \sim 0.5. \tag{5.4}$$

Therefore, flux changes by a factor of 4 between the orbits of Venus and Mars. Even in the case of a solar-like star the HZ will be restricted to a small circumstellar region of about 0.7 AU extension, corresponding to the difference between the semi-major axis of Mars and Venus.

In order for life to evolve to higher forms, the planet must be in the HZ continuously. The luminosity of solar like stars slowly increases during their main sequence evolution. The HZ slowly progresses outward due to stellar evolution. This leads to the definition of a continuously habitable zone (ContHZ):

"The ContHZ is the region in space where a planet remains habitable for some long time period τ_{hab}."

The evolution of intelligent life on Earth took $\sim$4 Gyr, and this must correspond to about τ_{hab}. Some other authors take smaller values such as $\tau_{\text{hab}} = 3\,\text{Gyr}$ or $\tau_{\text{hab}} = 1\,\text{Gyr}$, for the evolution of microbiological life. From these values it becomes clear that a planet must remain in a CHZ a considerable fraction of its lifetime in order for life to originate on its surface.

5.3.3 *Obliquity and habitability*

The obliquity of the rotation axis plays an important role in the climate and the atmospheric circulation of planets. The stabilizing effect of the Moon on the rotational axis of Earth is important for a relatively constant climate on Earth. Using a general circulation model of intermediate complexity, Planet Simulator (PlaSim), aquaplanets were modeled at varying obliquities between 0° and 90° [Snell *et al.*, 2017]. It turned out that:

- Planets with low obliquities have warm climate, the mean surface temperatures remain almost constant throughout the year.
- Planets with high obliquities have the mean global temperatures drop below freezing, and the planets experience strong seasons with large regions covered in ice at times throughout the year.

A planet is heated differently on its surface. If there is no obliquity, it is strongly heated near the equator and less heated near to the poles. The hot air rises and cools and sinks down near to the poles. This explains the pattern of Hadley cells (Fig. 5.5). These two regimes are the result of changes in the Hadley cells. Though this is a very simplified picture, in reality more than just one cell is found on Earth, the Hadley circulation explains the two regimes found above.

5.3.4 *Circumstellar HZ*

The inner and outer boundaries of the HZ are [Kasting *et al.*, 1993]:

$$r_o/r_i \sim [L(3.5)/L(1.0)]^{1/2}. \tag{5.5}$$

Here r_i denotes the inner boundary and r_o the outer boundary of a CHZ. $L(t)$ represents the luminosity after t billion years of a main sequence star of mass M.

For stars with different masses, we see that:

- r_o/r_i becomes smaller for less massive stars;
- It converges to $r_o = r_i$ for stars with $M \sim 0.83 \, M_\odot$.

  This would imply that low-mass cool stars (of spectral type later than K1) have no CHZ. The probability of finding a planet in such a narrow zone is quite low. This is now seen as less restrictive and HZ around late-type stars are believed to exist. Since less massive stars comprise the majority of stars in the universe, the chance of finding life elsewhere becomes higher.

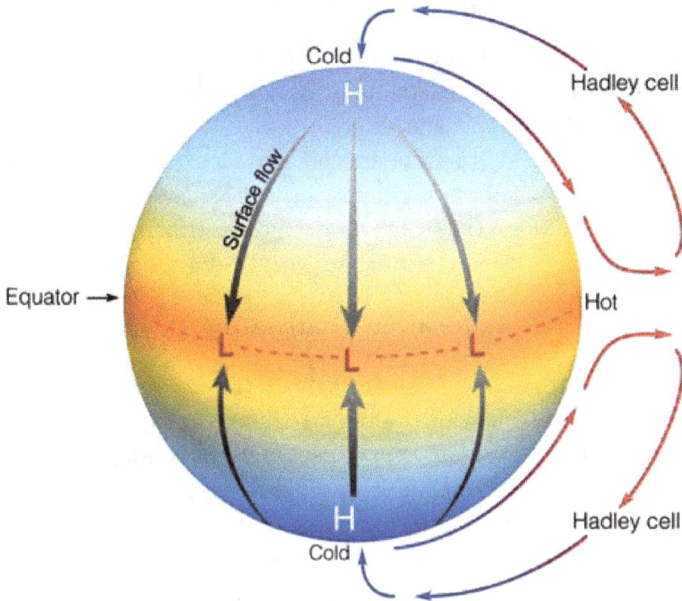

Fig. 5.5   Formation of a Hadley cell on a planet. Credit: SEAS Harvard.edu.

Up to now we have discussed radiation fluxes integrated over all wavelengths. For life, however, the more damaging part of radiation comes from short wavelengths like UV and X-rays.

Circumstellar habitable zones (CHZ) strongly depend on UV radiation.

- On Earth, UV radiation between 200 and 300 nm is very damaging to biological systems.
- On the other hand, it must be taken into account that UV radiation on the primitive Earth was one of the most important energy sources for the synthesis of biochemical compounds and was therefore essential to biogenesis.

In Kasting *et al.* [1993], circumstellar HZ around main sequence stars are studied. Stars later than F0 have main sequence lifetimes exceeding 2 Gyr and, thus, are potential candidates for harboring habitable planets. The authors argue that a log distance scale is probably the appropriate scale for this problem because the planets in our own solar system are spaced logarithmically. Moreover, the distance at which another star would be expected to form planets should be related to the star's mass.

Concerning late-type stars it can be stated that:

- F stars have larger and more extended HZ that are farther away from the parent star. Therefore, stellar activity should become less important for possible habitable planets.
- HZ around K and M stars are smaller and closer to the star; therefore, stellar activity becomes more important for planets in an HZ.
- Optimal candidates in the search for extraterrestrial life should be planets in an HZ around G stars and mid-to-early K stars.

5.3.5 CO_2 clouds and volcanic outgassing

As we have seen, there are several ways to define the outer boundary of the HZ in our solar system. All definitions require a rocky planet with liquid water on its surface. Assuming an atmosphere that contains N_2–CO_2–H_2O, the outer boundary of the HZ is at approximately 1.7 AU.

However, clouds have a strong impact on the climate of planetary atmospheres. The scattering greenhouse effect of CO_2-ice clouds is especially important.

For cool M stars, CO_2 clouds only provide about 6 K of additional greenhouse heating in the best-case scenario. On the other hand, the surface temperature for a planet around an F-type star can be increased by 30 K if carbon dioxide clouds are present [Kitzmann, 2017].

At larger distances, another process comes into play: in a planetary atmosphere containing CO_2, condensation and the scattering of this component starts because of the cool temperatures. This effect strongly diminishes the greenhouse effect that is normally dominant. It can be shown that a continuous outgassing of H_2 due to intense volcanism can strongly diminish this effect and the HZ becomes extended to a limit of about 2.4 AU [Ramirez and Kaltenegger, 2017].

The incident stellar flux that is needed to maintain liquid water surface temperatures at the outer edge of our solar system decreases from 0.3576 to 0.267 (25% decrease) and to 0.201 (44% decrease) with 5% and 30% H_2, respectively. Let us assume a hydrogen concentration of 50%. Then, for M–A stars, the stellar flux required to support the outer edge of the HZ reduces to 35–60% and the distances increase by 30–60%.

Thus, there is a warming effect of an H_2-rich planetary atmosphere. It should be noted, however, that H_2 escapes from earth-like planets, and therefore it must be continuously supplied by volcanism.

5.4 Evolution of HZ

5.4.1 *The faint young Sun problem*

The luminosity of the Sun has changed from 0.75 $L_\odot$ to its present value. At an age of only several hundred million years, the early Sun was fainter than the present Sun, so why there was water in liquid form even on early Earth, as we know from sedimentary deposits? A faint young Sun would result in a completely frozen world over the first 2 billion years in the history of our planet if all other parameters controlling Earth's climate had been the same. Yet there is ample evidence for the presence of liquid surface water and even life in the Archean (3.8–2.5 billion years before present). Therefore, some effect (or a combination of effects) must have been compensating for the faint young Sun. A wide range of possible solutions has been suggested and explored during the last decades, with most studies focusing on higher concentrations of atmospheric greenhouse gases like carbon dioxide, methane or ammonia. The faint young Sun paradox is reviewed e.g. in Feulner [2012].

As we have demonstrated in the short chapter about stellar evolution, nuclear fusion ignited in the core of the Sun when it has reached the main sequence in the HRD (this is also called zero main age sequence, ZAMS). This occurred 4.57 Gyr ago. The bolometric luminosity of the Sun (the solar luminosity integrated over all wavelengths) was about 30% lower than the present epoch. The long-term evolution of the bolometric solar luminosity $L(t)$ as a function of time t can be approximated by a simple formula

$$\frac{L(t)}{L_\odot} = \frac{1}{1 + \frac{2}{5}\left(1 - \frac{t}{t_\odot}\right)}, \tag{5.6}$$

where $L_\odot = 3.85 \times 10^{26}\,\text{W}$ (present day solar luminosity) and $t_\odot = 4.57\,\text{Gyr}$. This formula agrees very well with numerical calculations, but not for the first 200 million years. The increase of solar luminosity in time can be explained by the virial theorem: the Sun generates energy by nuclear fusion of hydrogen to helium in its core. Over time, helium nuclei accumulate, increasing the mean molecular weight within the core.

According to the virial theorem,

$$2E_\text{kin} = |E_\text{pot}|, \tag{5.7}$$

twice the total kinetic energy is equal to the absolute value of the potential energy. Therefore, the Sun's core contracts and heats up to keep the star stable, resulting in a higher energy conversion rate (which is strongly dependent on temperature) and hence a higher luminosity.

Solar variability (and thus ultraviolet luminosity) was higher in the past due to a steady decrease in magnetic activity over time caused by the gradual slowing of the Sun's rotation that ultimately drives the dynamo. The decrease of solar rotation can be approximated by

$$\Omega_\odot \propto t^{-0.6}. \tag{5.8}$$

Also the solar wind was stronger for the young Sun. This had a strong influence on the early Earth's atmosphere. Volatiles and water escaped. This process was even more enhanced because the early Earth's magnetic field strength was only about 50–70% of its present value. We can easily estimate the influence of a changing solar constant $S = 1360\,\text{W/m}^2$ on the incoming flux F on Earth:

$$\Delta F = \Delta S_0 (1 - A)/4. \tag{5.9}$$

A is the Earth's albedo (the current value is 0.3).

Let us assume that the solar constant $S = 0.75 S_0$, where S_0 is the present day value. Then $\Delta S = 360$ and $\Delta F = 63\,\text{W/m}^2$. For comparison, the net anthropogenic radiative forcing in 2005 is estimated to be $\sim 1.6\,\text{Wm}^{-2}$.

There is an ice–albedo feedback. Ice has a higher albedo, and more ice means that when more radiation is reflected back into space, the cooling will be enhanced. This is also known as runaway glaciation or snowball Earth. Models show that such an effect is expected as soon as a threshold of about 15–18% decrease in solar luminosity is reached. To return to a milder climate, high concentrations of greenhouse gases (CO_2 and others) are required. For Earth it is estimated that e.g. a 3 bar CO_2 atmosphere would be sufficient for maintaining a warm climate. A higher concentration of such gases in the early Earth's evolution could be explained by: intense volcanism or meteorite impacts. This is discussed in Zahnle [2008]. For Mars, impact erosion during the early bombardment may explain its thin atmosphere.

Also the effect of geothermal heating was more important at that time and the oceans would not have been frozen completely.

Let us repeat the formation of Earth and the early evolution:

- Formation: by gravitational accretion of smaller bodies (planetesimals) formed in the nebula surrounding the young Sun.

- Origin of the Moon: the large impact forming the Moon occurred after 50 Myr toward the end of the accretion period.
- After this event, the Earth was enshrouded in rock vapor for 1,000 years.
- The surface of the Earth had a magma ocean for a few million years because of a strong greenhouse effect (caused by large amounts of carbon dioxide and water vapor degassing from the mantle) and tidal heating by the still tightly orbiting Moon.
- Then, the crust solidified and a hot water ocean with temperatures of about 500 K formed under a dense atmosphere containing about 100 bar of carbon dioxide.
- The carbon dioxide in the atmosphere was then subducted into the mantle over timescales of 10^{7-8} years, before the Late Heavy Bombardment (4.0–3.8 Gyr).
- Geologic evidence for liquid surface water during the Archean is mostly based on sedimentary rock laid down in a variety of aqueous conditions up to 3.5 Gyr and possibly as early as 3.8 Gyr, and there is no evidence for widespread glaciations during the entire Archean.

In summary, a solution of the faint young Sun paradox for Earth is possibly a combination of several effects:

- Enhanced greenhouse gases (ammonia, methane, carbon dioxide, etc.).
- Faster rotation of early Earth.
- Different continental configuration and distribution.

On the other hand, it is evident that an increasing solar luminosity means that HZ tend to move outward with time because the Sun, like all main sequence stars, has become brighter during its evolution.

5.4.2 *The evolution of circumstellar habitable zones*

We have already discussed how stars evolve and that their luminosity increases, as it is well known for the Sun. Since the evolution of life on Earth took about 4 billion years, a ContHZ was required.

The equilibrium temperature T_P on a planet depends on the effective stellar flux S_{eff}, the albedo A and a, which is a conversion factor between centimeters and astronomical units:

$$T_P = \left(\frac{S_{eff}(1 - A)L_\odot}{16\pi\sigma a^2} \right)^{1/4}. \tag{5.10}$$

We can assume the inner boundary at $T_i = 269\,\text{K}$. This value guarantees that the surface temperature remains below the critical temperature of water (647 K). A planet with an atmosphere at an equilibrium temperature greater than 270 K either

- has a surface temperature below 647 K but no liquid water on its surface,
- or a considerable amount of water but a surface temperature above 1400 K.

The planet would be uninhabitable for life.

The definition of the outer boundary is complicated. Stable liquid water may exist under a dense CO_2 atmosphere. According to the work of Valle *et al.* [2014], analytical expressions for the HZ vary and its evolution can be approximated. We give a few examples from that paper:

$$Z_1 = \log Z, \tag{5.11}$$
$$M_1 = \log M, \tag{5.12}$$
$$K_1 = \log(\sqrt{(1-A)/0.7})/T_{169}^2, \tag{5.13}$$
$$T_{169} = T_0/(169\,\text{K}), \tag{5.14}$$
$$K_2 = \log\sqrt{(1-A)/0.7}. \tag{5.15}$$

Z denotes the metallicity of the star. Assuming that d_m, t_m, W, R_i, R_o depend on the mass of the host star M, the metallicity Z and the helium abundance Y, some analytical expressions can be given. For the distance d_m at which the HZ has the longest duration, the following analytical formula can be given:

$$\log d_m(\text{AU}) = K_1 + (M * Z_1 + M^2 * Z_1^2 + M^3 * Z_1^3) * Y. \tag{5.16}$$

The operator $*$ is defined by

$$A * B = A + B + AB. \tag{5.17}$$

The chemical inputs are Y and Z. For example, the boundaries of the CHZ are 0.8–3.3 AU for a duration of 1 Gyr and 0.9–2.8 AU for a duration of 3 Gyr for a host star of $1.0\,M_\odot$, $Z = 0.017$, $Y = 0.26$.

The longest duration t_m in the HZ in Gyr is given by

$$\log t_m(\text{Gyr}) = \left(\frac{1}{M} + M_1 * Z_1 * Y + M_1^2 * Z_1^2\right) * (T_{169} + T_{169}^2). \tag{5.18}$$

The width of the HZ is given by

$$\log W(\text{AU}) = K_2 + (M * Z_1 + M^2 * Z_1^2 + Z_1^3) * YT_{169}, \qquad (5.19)$$

the scaling $K_2 = \log \sqrt{(1-A)/0.7}$. For the inner boundary R_i and the outer boundary R_o, the following values are obtained:

$$\log R_i(\text{AU}) = K_2 + (M * Z_1 + M^2 * Z_1^2 + M^3 * Z_1^3) * Y,$$

$$\log R_o(\text{AU}) = K_2 + (M * Z_1 + M^2 * Z_1^2 + M^3 * Z_1^3) * Y * T_{169} + T_{169}^2.$$

5.4.3 *Circumbinary habitable zones*

A large percentage of all G–M stars in the solar neighborhood are expected to be part of binary or multiple stellar systems. Furthermore, current statistics of the Kepler candidates exoplanet population suggest that about 46% of all planets discovered so far reside in multiplanet system. Are habitable planets likely to be discovered in such environments? As current exoplanet statistics predicts that the chances are higher to find new worlds in systems that are already known to have planets, four known extrasolar planetary systems in tight binaries were investigated by Funk *et al.* [2015] in order to determine their capacity to host additional habitable terrestrial planets. Those systems are Gliese 86, γ Cephei, HD 41004 and HD 196885.

Binary systems in which a planet orbits close to one star, are called S-systems. Habitable planets in such systems were investigated in Kaltenegger and Haghighipour [2013]. In P-systems, the planet orbits both stars [Haghighipour and Kaltenegger, 2013].

In the case of γ Ceph, only the M dwarf companion could host additional, potentially habitable worlds. No stable and no potentially habitable regions were found around HD 196885 A. HD 196885 B can be considered a slightly more promising target in the search for Earth-twins. Gliese 86 A turned out to be a very good candidate, assuming that the system's history has not been excessively violent. For HD 41004, stable orbits for habitable planets were found. A more detailed investigation shows that for some initial conditions, stable planetary motion is possible in the HZ of HD 41004 A. In spite of the massive companion HD 41004 Bb, it was found that HD 41004 B too could host additional habitable worlds [Funk *et al.*, 2015].

The effect of the stellar flux on the HZ is becoming an increasingly important property of exoplanetary systems as more planets are discovered within the HZ. Recently, the Kepler mission has uncovered circumbinary planets whose HZs can be relatively complex due to the flux from binary host stars. In Kane and Hinkel [2013], HZ boundaries for circumbinary systems are derived. These depend on the stellar masses, separation and time dependence when accounting for orbital motion and the orbit of the planet.

5.5 Circumplanetary habitable zone

So far, HZs have been discussed around a central star. If an HZ is defined by the presence of liquid water, we must also include the interiors of giant planets and the ice covered Galilean Moons of Jupiter, the Martian subsurface and maybe even other places.

Therefore, the search for life must not be restricted to the study of CHZ. However, it seems that life requires much more than just liquid water, i.e. an energy source.

5.5.1 *Heating processes*

Such an energy source can be provided by mechanisms other than radiation from a central star in the cases of satellites of planets.

One possible mechanism in connection was already discussed: the tidal heating. An overview can be found in Reynolds *et al.* [1987].

Tidal forces result in deforming an object orbiting a larger one. Such a continuous compression produces heat.

Tidal heating occurs at a rate of

$$F_{\text{tid}} = (9/19)\rho^2 n^5 R^5 e^2 \frac{1}{\mu Q}, \tag{5.20}$$

where n is the moon's mean motion about the planet, μ is the moon's rigidity ($6.5 \times 10^{11}\,\mathrm{dyn\,cm^{-2}}$ for Io), e the eccentricity of the moon from which the tidal surface heating of Io can be calculated $\sim 41\,\mathrm{erg\,cm^{-2}\,s^{-1}}$.

All these considerations led to the conclusion that a $0.12\,M_\oplus$ moon ($M_\oplus$ being the mass of Earth) in an Io-like orbital resonance and possessing a Ganymede-like magnetic field could remain habitable for several billions of years. Therefore, systems belonging to 47 UMa and 16 Cyg B could have possible moons around their giant planets as candidates for extraterrestrial life.

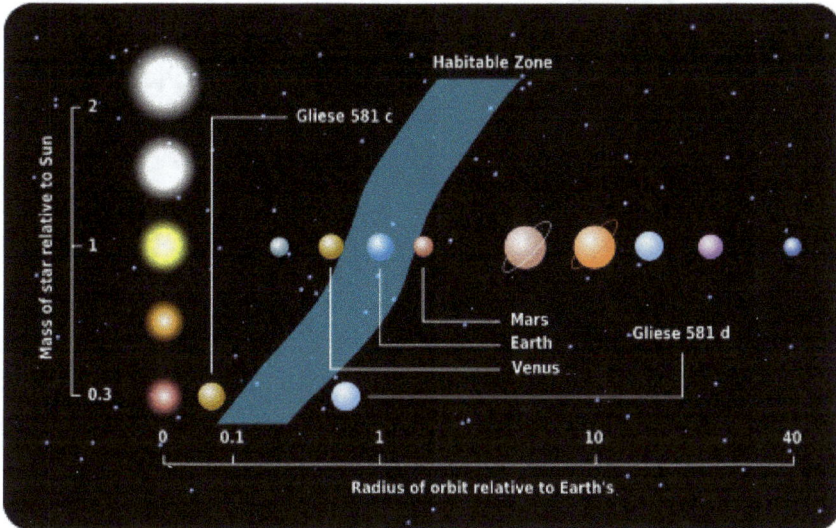

Fig. 5.6 Comparison of the CHZ in our solar system and the Gliese system. On the ordinate stars of different masses are shown. The HZ is shifted to larger distances from the central star for larger stellar masses. Credit: http://its-interesting.com.

5.5.2 *Tidal heating: Io*

Io is one of the four Galilean satellites of Jupiter that were first seen by Galilei in 1609.

On the surface of Io, 400 active volcanoes and more than 150 mountains have been detected, as well as volcanic plumes extending up to 500 km. The tidal forces of Jupiter on Io are 6,000 times stronger than those of the Moon on Earth. There are also tidal forces caused by the two other Galilean satellites *Europa* and *Ganymede* (these are comparable to the tidal force of the Moon). Furthermore, the strength of these forces varies because the orbit of Io is elliptical. The variation of the tidal forces of Jupiter due to Io's elliptical orbit is 1,000 times the strength of the tidal forces of the Moon.

On Earth, the tidal force causes deformations of the whole Earth's crust between 20 and 30 cm. On Io, these deformations can reach up to 300 m.

To summarize, Io is perhaps the most geological active object in the solar system because of

- Tidal heating from Jupiter;
- Heating effects due to resonances with other satellites of Jupiter.

Fig. 5.7   The galactic HZ in a typical spiral galaxy.

5.6   Galactic habitability

5.6.1   *Galactic habitable zone*

In a typical galaxy there is a metal gradient. Near the center, the metal content becomes higher. The star density is higher there, with more stars evolving into supernovae and the interstellar matter becoming enriched by metals.

A planetary system must therefore be located close enough to the galactic center so that a sufficiently high level of metals could exist. Rocky planets could only have formed under these conditions. The necessity of heavier elements, such as carbon, for the formation of complex molecules for life is also evident.

Although planetary systems can only evolve within a certain distance from the galactic center, they should not be too close to the galactic center or else they risk perturbations by passing stars. Such perturbations will trigger hazards from comets moving into the inner planetary systems. Near the galactic center, outbursts from supernovae and from the supermassive black hole at the center could cause strongly enhanced short-wavelength radiation bursts destroying the complex molecules needed for life.

Observations of extrasolar planetary systems seem to show a trend that when the metallicity of the star becomes too high, the number of massive planets orbiting close to the stars increases; those massive planets can destroy Earth-sized objects.

It was shown by Sundin [2006] that, in a barred galaxy, the GHZ is more complicated to define. The central bar can change stellar orbits. This is also valid for our Milky Way galaxy, since it is a barred galaxy. For the dynamics of such objects, two effects are important:

- Stars in the bar: their motion describes complicated orbits.
- Stars outside the bar: these bars will influence the orbits of stars in the whole galaxy. Stars passing close to the bar can either gain or lose angular momentum, due to a positive or negative torque by the bar. Some stars will therefore be captured by the bar while some stars may eventually reach the escape velocity from the galaxy.

5.6.2 *The evolving galactic habitable zone*

In the early stages of galaxy evolution, the heavy elements needed to form terrestrial planets were only present near the center of the galaxy. This was due to the fact that the concentration of stars there was largest and some stars had already evolved and became supernovae, which enriched the interstellar medium with heavier elements.

However, this was not a safe environment and perhaps no continuous galactic HZ existed.

Gradually, the heavy elements spread through the galaxy and terrestrial planets formed at greater and safer distances from the galactic center. The HZ appeared about 8 billion years ago at a certain distance from the galactic center. An annular GHZ formed. In our galaxy, the galactic HZ is an annular region 7–9 kpc[3] from the galactic center. This zone becomes larger with time and is composed of stars formed between 4 and 8 billion years ago. This was described by Lineweaver *et al.* [2004].

---

[3]pc denotes parsec, 1 pc = 3.26 light years or 32.6 × 10^{12} km.

75% of the stars in the galactic HZ are older than the Sun. They are 1 Gyr older than the Earth. Therefore, our civilization might belong to the "young generation" of galactic civilizations [Lineweaver, 2007].

The galactic chemical evolution can substantially influence the creation of habitable planets. Metallicity plays a crucial part. There is observational evidence that the lifetimes of circumstellar disks are shorter at lower metallicities. This can be explained by greater susceptibility to photoevaporation.

The disk lifetime can be estimated as

$$t_{\text{disk}} = 2\text{Myr} \left(\frac{Z}{Z_\odot} \right)^{0.77(4-2p)(5-2p)}, \tag{5.21}$$

where p is the power-law exponent of the disk surface density profile (e.g. $p = 0.9$); q is the power-law exponent of the disk temperature profile (e.g. $q = 0.6$).

The fiducial temperature profile depends on the distance from the central star

$$T(r) = 200\,\text{K} \left(\frac{r}{1\,\text{AU}} \right)^{-q}. \tag{5.22}$$

In fact, the model of Johnson & Li Johnson and Li [2012] showed that the first Earth-like planets likely formed from circumstellar disks with metallicities $Z \leq 0.1 Z_\odot$.

Galactic chemical evolution models can be useful for studying the galactic HZ in different systems [Spitoni *et al.*, 2014]. Detailed chemical evolution models including radial gas flows were applied in order to study the galactic HZ in our galaxy and in M31 (Andromeda galaxy). The results were compared to the relative galactic HZ found with "classical" (independent ring) models, where no gas inflows were included. For both the Milky Way and Andromeda, the main effect of the gas radial inflows is to enhance the number of stars hosting a habitable planet. These results were obtained by taking the supernova destruction processes into account.

The main results of this study can be summarized as:

- Milky Way galaxy: the maximum number of stars hosting habitable planets is at 8 kpc from the Galactic Centre, and the model with radial flows predicts a number that is 38% larger than what was predicted by the classical model.
- For the Andromeda galaxy, the maximum number of stars with habitable planets is at 16 kpc from the centre and, in the case of radial flows, this number is 10% larger than the number of the stars predicted by the classical model.

The total number of stars formed at a certain time t and galactocentric distance R hosting Earth-like planet with life is given as

$$N_{\text{life}}(R, t) = \text{PGHZ}(R, t) \times N_{\text{tot}}(R, t) \qquad (5.23)$$

where $\text{PGHZ}(R, t)$ is the fraction of all stars having Earths (but no gas giant planets) that survived supernova explosions as a function of the galactic radius and time.

5.7 Potentially habitable bodies

As we have shown, the definition of a HZ depends on whether we consider a planetary system or a whole galaxy. Therefore, it seems easier to define habitable bodies. According to Lammer *et al.* [2009] four classes can be defined.

5.7.1 *Classes of habitable bodies*

- Class I: bodies on which stellar and geophysical conditions allow Earth-analog planets to evolve so that complex multicellular life forms may originate;
- Class II: bodies on which life may evolve but, due to stellar and geophysical conditions that are different from the class I habitats, the planets rather evolve toward Venus- or Mars-type worlds where complex lifeforms may not develop;
- Class III: bodies where subsurface water oceans exist that interact directly with a silicate-rich core, like Europa;
- Class IV: bodies in which there are large liquid water layers between two ice layers or liquids above ice (also called super-Ganymedes or ocean planets).

5.7.2 *Distribution of habitable bodies*

The distribution of habitable bodies is, at present, difficult to estimate. Class I objects, Earth-analogs, are at the onset of detection. Several super-Earth objects have been found already. An example is given in Fig. 5.8. COROT 7b has a diameter 1.53 times that of Earth. It orbits the host star in only 20 hours and is very close to it (about 1/20 the distance between Mercury and the Sun).

For example Anglada-Escudé *et al.* [2012] report on a planetary system around the nearby M dwarf GJ 667C with at least one super-Earth in its

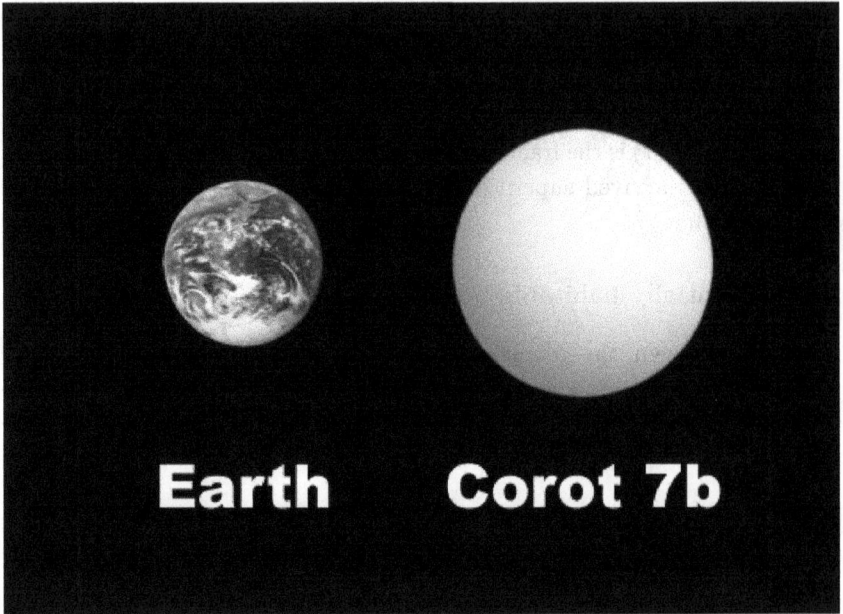

Fig. 5.8   A size comparison of Corot 7b and Earth. Source: Wikimedia.

HZ. GJ 1214b, the 6.55 Earth-mass transiting planet recently discovered by the MEarth team, has a mean density of ∼35% that of the Earth. It is thought that this planet is either a mini-Neptune, consisting of a rocky core with a thick, hydrogen-rich atmosphere, or a planet with a composition dominated by water [de Mooij *et al.*, 2012]. The habitability of super-Earth planets around main-sequence stars, including red giant branch evolution was studied by Cuntz *et al.* [2012].

As it was pointed out already, several protection mechanisms must be provided for a planet to be habitable. One of these is an intrinsic magnetic field based on a dynamo. The magnetic fields of planets (like stars) are related to their rotation. A magnetic field of long duration depends mainly on two factors:

- planetary mass,
- rotation rate.

It seems that low-mass super-Earths ($M \leq 2\ M_\oplus$) develop intense surface magnetic fields but their lifetimes are limited to 2–4 Gyrs for rotational periods longer than 1–4 days [Zuluaga and Cuartas, 2012].

Chapter 6

Stellar Activity and Habitability

After having discussed all different aspects of stellar evolution, planets, planetary atmospheres and planetary and atmospheric evolution, we will focus on stellar activity.

In this chapter we will summarize and discuss all the effects of stellar activity on the habitability of the planets surrounding them.

6.1 Stellar mass loss

In 1924, Milne pointed out that stars with high temperatures can have strong winds because the radiation pressure can eject atoms/ions from their atmospheres.

By these processes stars are constantly losing mass; the mass loss rate depends on

- stellar age,
- location of a star in the HRD.

The emitted particles can reach the planets and different processes cause erosion or some chemical reactions in the planetary atmospheres; therefore, habitability on a planet can get lost in extreme cases.

6.1.1 *Stellar winds*

A typical observational signature of extremely strong winds are P Cygni profiles.[1] Such a line profile can be explained by an expanding outflow as it is shown in Fig. 6.1.

---

[1] Named after the star P Cygni.

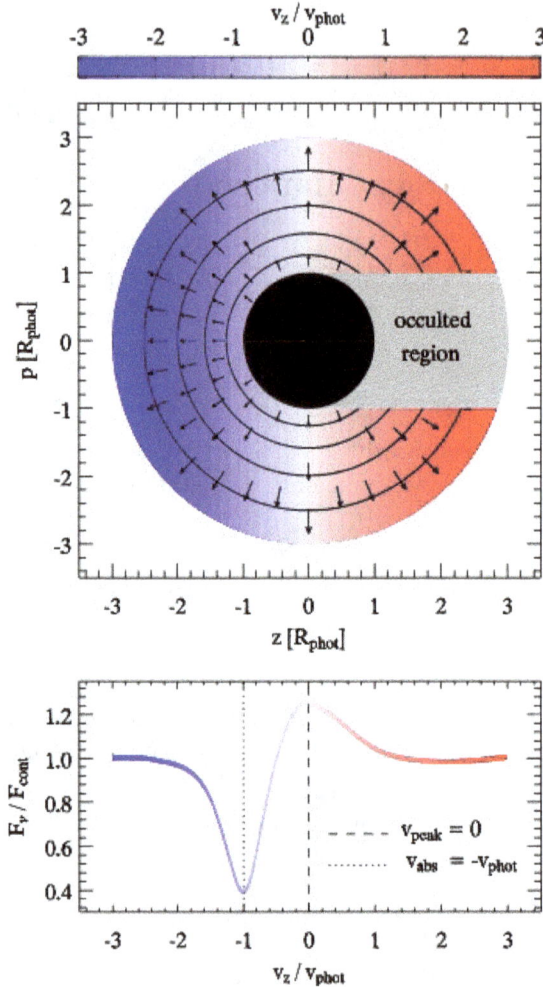

Fig. 6.1 Explanation of a typical P Cygni line profile. The observer is on the left and sees the expanding shell where absorption occurs. Since the shell approaches the observer the absorption lines are blueshifted with respect to the emission lines. Credit: St. Blondin.

In Fig. 6.2 the different types of stellar winds are shown for different types of stars in the HRD. Basically, there are three types of stellar winds:

- hot, solar-type stellar winds,
- cool, dense slow winds,
- radiatively driven winds for hot stars with no convection zone on their surface.

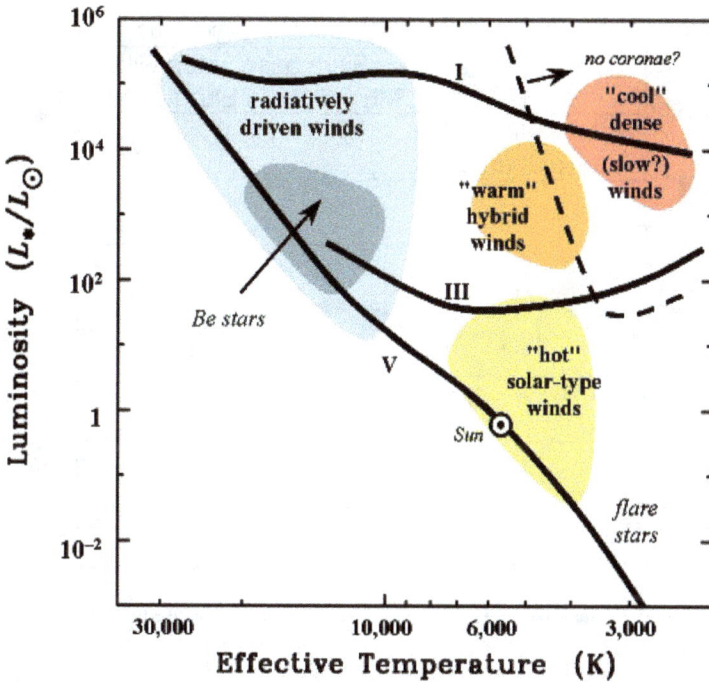

Fig 6.2 Different types of stellar wind depending on spectral type of stars. Credit.
Harvard–Smithsonian CfA.

There is also a hybrid case, as is shown in the figure. Stars with extended
coronae (high temperature above 1 million K) accelerate hot, solar-type
winds. The mass loss is low, $<10^{-12} M_\odot/\mathrm{yr}$. The cooler, evolved stars
exhibit chromospheres with lower temperatures (around 10,000 K) and
they have winds with lower speeds; however, the mass loss rates are larger
($10^{-7} M_\odot/\mathrm{yr}$).

An interesting study on the 500 million years old star π^1 UMa was made
by Wood *et al.* [2014]. For that case, the Ly α absorption observed with
the HST spectroscope and a stellar wind only half as strong as the solar
wind was found. This suggests that the Sun and solar-like stars did not
have particularly strong coronal winds in their youth.

6.1.2 Stellar flares and CMEs

As was already discussed in relation to the Sun, there seems to be a
correlation between strong flares and CMEs. Such a relation, however, is

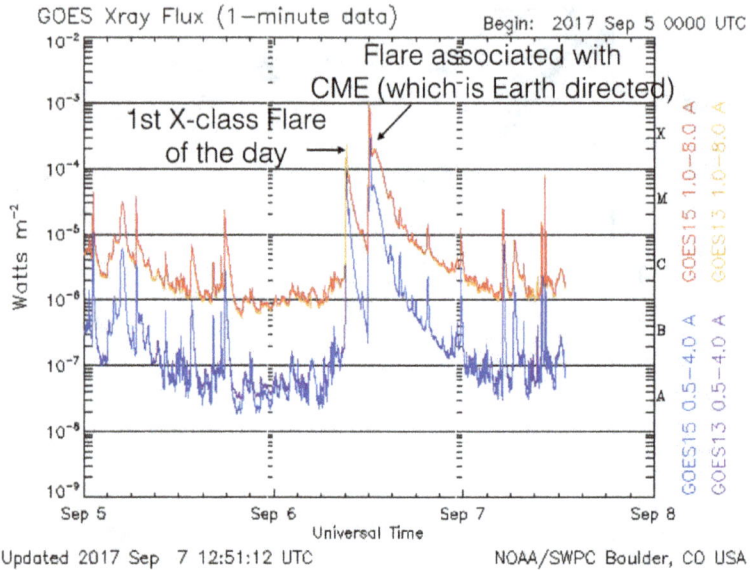

Fig. 6.3 The X-class flares in September 2017 were associated with CMEs. Credit: NASA.

difficult to establish for stars, and the exact contribution of stellar CMEs to mass loss over stellar evolution is still unknown. It is extremely important to know this relation for planetary habitability. Stellar mass loss influences on stellar evolution (loss of angular momentum) and also on planetary atmospheres (especially in the case of close planets with weak magnetic fields). In Odert *et al.* [2017] an empirical model is presented in which solar flare–CME relationships are combined with stellar flare rates to estimate the CME activity of young Sun-like and late-type main-sequence stars.

An example of a solar flare outburst associated with a CME is shown in Fig. 6.3.

Some characteristics like energy release and occurrence rate of flares on different stars are shown in Fig. 6.4. In Maehara *et al.* [2012], 365 superflares were studied on solar-type G stars using data from the Kepler mission. The maximum energy of the flare is not correlated with the stellar rotation period, but the data suggest that superflares occur more frequently on rapidly rotating stars. It has been proposed that hot Jupiters may be important in the generation of superflares on solar-type stars; however, none have been discovered around the stars that have been studied, indicating that hot Jupiters associated with superflares are rare.

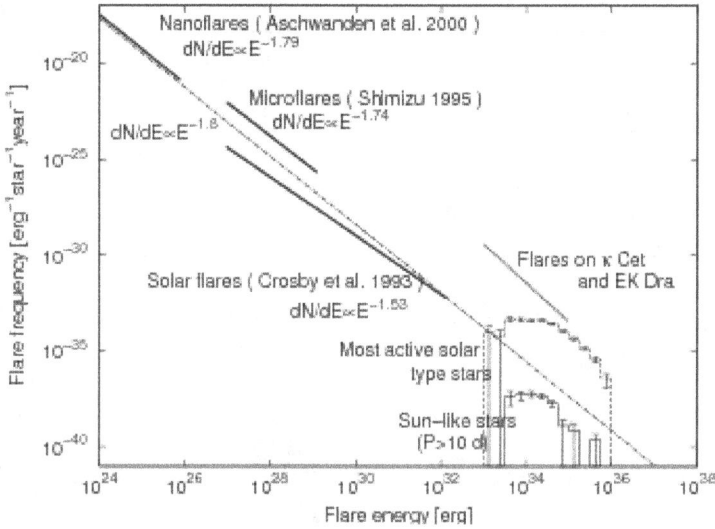

Fig. 6.4 Flare energies and occurrence rates on Sun-like stars. Adapted from Maehara *et al.* [2012].

Table 6.1. Observed and predicted flare parameters for some stars and the Sun.

Star	Sp. Type	L_x ergs s^{-1}	EM cm^{-3}	L_x/L_{opt} observed	L_x/L_{opt} predicted
AT Mic	dM4.5e	1.6×10^{31}	1.4×10^{54}	>4	<0.73
AD Leo	dM3.5 e	4.6×10^{30}	4.0×10^{53}		
UV Cet	dM6e	2.0×10^{30}	1.2×10^{53}	0.04	0.029
YZ CMi	dM4.5e	8×10^{30}	5×10^{53}	0.03	0.066
Prox Cen	dM5e	$0.7–1.8 \times 10^{30}$	4.6×10^{53}	>2	<0.03
Sun	G2V	4×10^{27}	6×10^{49}	2	0.030

In Table 6.1 some observed and predicted flare parameters for some stars are given.

Finally, we offer a comparison between large solar and stellar flares in Table 6.2.

CMEs could contribute to an expansion of the habitable zone around low-mass stars by the supply of UV radiation (Oishi *et al.* [2017]). CME activity of a star may vary depending on stellar age, stellar spectral type and the orbital distance of a planet. Because of relatively short range of propagation of majority of CMEs, they impact most strongly

Table 6.2.   Comparing large solar and stellar flares.

Object	Energy	Max. duration	Intensity increase (vis.)	Intensity increase (X-ray)
Sun	10^{32} ergs	∼5 hours	1.0027	6000
young stars	10^{36} ergs	∼1 day	small	50
single stars	10^{35} ergs	several days	1000	500
binary stars	10^{38} ergs	∼1 week	10	120

the magnetospheres and atmospheres of close-orbit (<0.1 AU) exoplanets [Khodachenko *et al.*, 2014, 2007].

6.2   Planetary atmospheres

Before discussing the influences of stellar activity on planetary atmospheres, we give some remarks about their physics.

6.2.1   *Stars versus planets*

As we have seen in the chapter about stellar structure and evolution, stars are determined by their mass, initial chemical composition and age. This is also known as the Voigt–Russel theorem. The Voigt–Russel theorem explains why various stellar parameters are well correlated during stellar evolution, i.e. on the main sequence. A well-known example is the correlation between temperature and mass. Stellar parameters are closely correlated and, if one is known, others can be deduced.

In that sense, however, planets are more complex. Given only the mass of a planet we can not say whether it is a gas giant, a rocky object or an icy planet. Planetary mass, composition, temperature and other parameters are only loosely correlated; in some cases, there seems to be no correlation at all. The orbits of the solar system planets appear to be almost circular; however, there are many known cases of exoplanets with orbits of high eccentricity and, in contrast to the solar system, a large amount of close-in exoplanets ($a < 0.05$ AU) has been found.

An example of a planet with extreme eccentricity is HD 20782b. The semi-major axis is about 1.36 AU, the periastron at 0.10 AU and the apastron at 2.62 AU. Therefore, the eccentricity is 0.97. It is in the constellation of Fornax, at a distance of 117 ly. The difference between the periastron and apastron leads to extreme temperature changes on that planet.

The planet with the smallest orbit is PSR J1719-1438 b. It orbits a millisecond pulsar at a distance of 0.004 AU. This planet seems to be extremely dense and is probably composed of crystalline carbon, much denser than diamond. It has been suggested that, in this case, a star transformed into a planet in a millisecond pulsar binary [Bailes *et al.*, 2011].

This also leads to the question of whether the solar system is unique or belongs to the common systems of stars with planets. We cite the early work of Aitken [1938]: *Our conclusion then must be that the development of a solar system is rather an exceptional event, but that is no reason for believing our system to be unique. Put the odds against any star being the central body in a system like our own at a million to one, if you please; that would still provide 30,000 solar systems in our own stellar system.* Today, we know that planetary systems are quite common. However, most of the detected systems are completely different from our own system. Therefore, it is hard to answer whether or not our solar system is unique because of observational constraints. It is much easier to detect giant planets close to their parent stars than Earth-like planets.

Concerning habitability, the physical characteristics of a planet, the composition of its atmosphere, the existence of a magnetosphere, its orbital characteristics and, the parameters and activity of its host star are essential.

6.2.2 *Detection of exoplanetary atmospheres*

The stellar irradiation on a planet depends on several factors:

- Distance from its hosting star.
- Eccentricity of the planetary orbit; 60% of detected exoplanets move on high eccentric orbits.
- Rotation period; for planets orbiting very close to their stellar companion, in fact, gravitational interaction with their host star may result in the orbit circularization and the synchronization of the rotation and revolution periods.
- Axial tilt. In the solar system, all planets have an axial tilt $<27°$ with the exception of Venus and Uranus.

The surface temperature, which is the relevant parameter for habitability, is influenced by stellar irradiation and moreover by

- the composition of the planet's atmosphere,
- internal heating.

At present, three techniques can be used to measure the atmospheres of exoplanets:

(1) the transit method,
(2) high-resolution Doppler spectroscopy, and
(3) direct imaging.

One of the most successful methods in detecting exoplanetary atmospheres is the transit method. When the planet transits in front of the host star, part of the star light transverses through the day–night terminator region of the planetary atmosphere. We observe a transmission spectrum. This spectrum can be obtained from a subtraction of the in-transit spectrum from the out-of-transit spectrum. Several absorption features that can be observed are from the planetary atmosphere. Transmission spectra probe the chemical composition and temperature structure at the day–night terminator of the planet. The planet is at full phase before being occulted by the star (secondary eclipse). In that case, the thermal emission and reflection spectrum from the planet are observed along with the stellar spectrum. Again, by subtracting the stellar spectrum, the signatures in the planet's atmosphere can be observed. The stellar spectrum itself can be observed, when the planet is totally occulted by the host star. Secondary eclipse spectra can be used to probe composition and thermal structure of the dayside atmosphere of the planet.

In Fig. 6.5 the primary and secondary eclipse as well as the phase variation of a transiting exoplanet is shown.

In Fig. 6.6 the lightcurve variation due to exoplanet HAT-P-7b is shown. This is a hot Jupiter at a distance of only 5.7 million km from its host star. It orbits its host star in only 2.2 days. The host star is at a distance of 1,040 ly. The host star has a surface temperature of 6,350 K, a mass of 1.1 $M_\odot$ and a radius of 1.5 $R_\odot$. Signs of powerful changing winds have been detected on the planet 16 times larger than Earth. This discovery was made by monitoring the light being reflected from the atmosphere of HAT-P-7b, and identifying changes in this light, showing that the brightest point of the planet shifts its position. This shift is caused by an equatorial jet with dramatically variable wind-speeds, pushing vast amounts of clouds across the planet. The clouds themselves are probably visually stunning; they are made of up corundum, the mineral that forms rubies and sapphires.

One side of the planet always faces the star, because it is tidally locked, and that side remains much hotter than the other. The day-side average temperature on HAT-P-7 is 2,860 K.

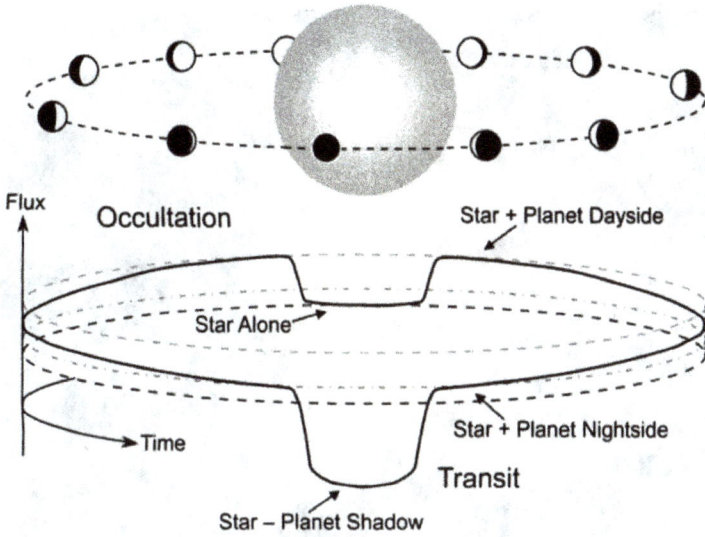

Fig. 6.5 A transiting exoplanet. Note the primary and secondary eclipse as well as the behavior of the phase. Credit: Joshua Winn.

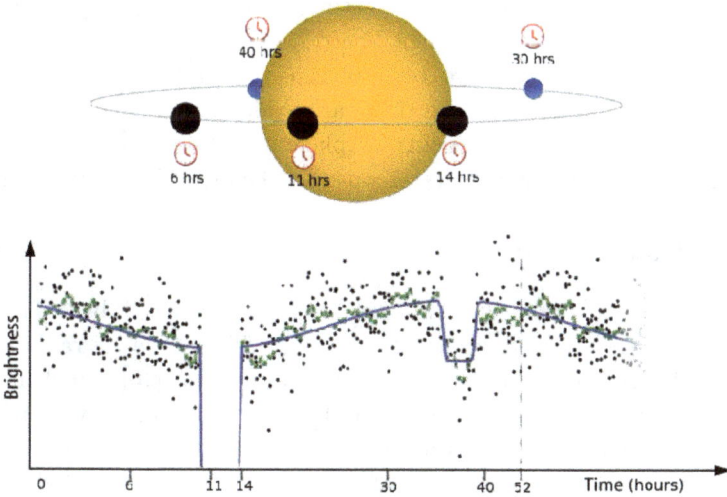

Fig. 6.6 The primary and secondary eclipses of exoplanet HAT-P-7b. Credit: W. Borucki & others/Science.

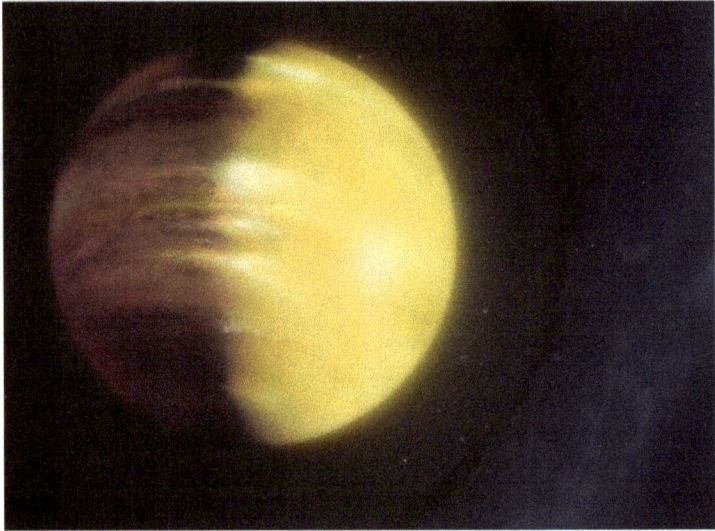

Fig. 6.7 An artist's conception of exoplanet HAT-P-7b. Credit: University of Warwick/Ronald Warmington.

An artist's conception of this planet with its unusually bright clouds is shown in Fig. 6.7.

Transmission spectra have been obtained from UV to IR. Thermal emission spectra predominate in the near to mid IR.

Other techniques for observing exoplanetary atmospheric features are by high-resolution Doppler spectroscopy and the direct imaging of planets.

6.2.3 *Photochemistry*

Planets, especially hot Jupiters, are strongly irradiated by their host stars. They are exposed to a high UV-flux. This type of radiation is absorbed in the upper atmosphere and the result is a rich photochemistry. Parent molecules are transported from deeper atmospheric layers and are photodissociated in upper layers, producing radicals that react to form new species. The photochemistry of gas planets like Jupiter and Saturn have been studied in the past. The photochemical processes depend on the planet's temperature:

- Cooler planet: CH_4 and NH_3 are present and they are active photochemical precursors.
- Hotter planet: CO and N_2 are present; these are less active precursors.

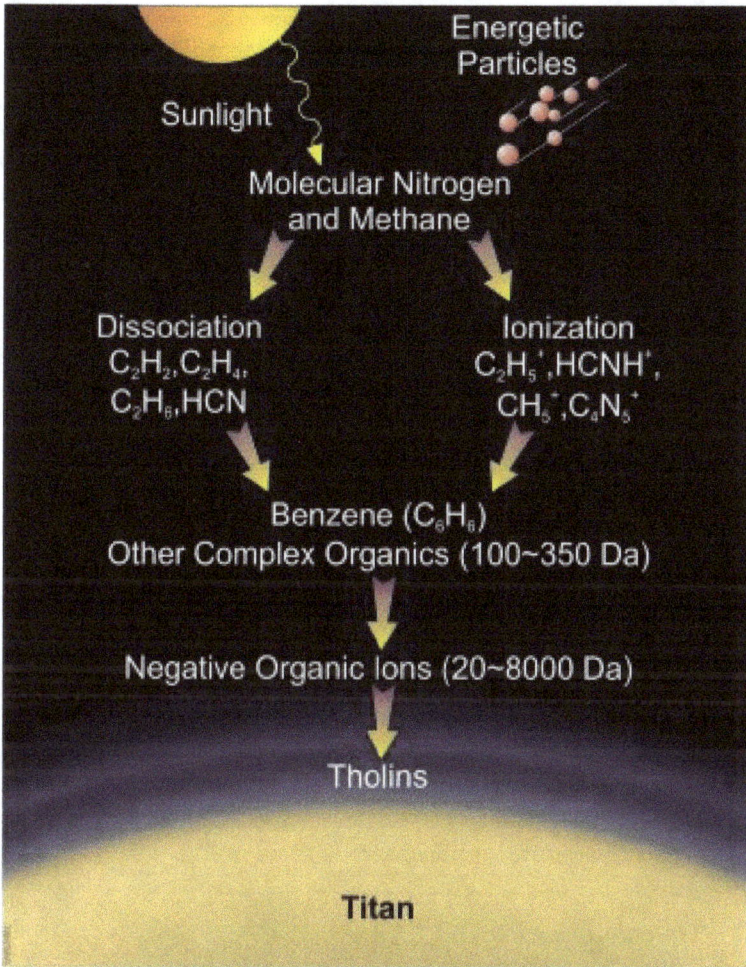

Tholin formation in Titan's upper atmosphere

Fig. 6.8   The Tholin formation process. Credit: Southwest Research Institute.

In Fig. 6.8 the formation of organic tholins in Titan's atmosphere is shown schematically.

Let us consider the photochemistry of Jupiter and Saturn. For both, the composition of the stratosphere is mainly controlled by disequilibrium chemistry initiated by methane photolysis. Due to the general similarities in stratospheric temperatures ($\sim$165 versus $\sim$140 K), methane abundances

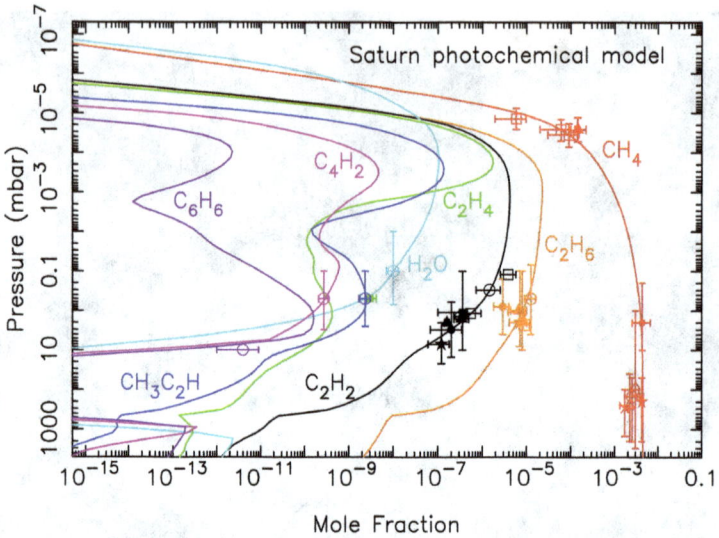

Fig. 6.9   Photochemistry in Saturn's atmosphere. Adapted from Moses, LPI.

($\sim 1.8 \times 10^{-3}$ versus $\leq 4.5 \times 10^{-3}$ mole fraction) and overall composition (major gases H_2, He, and CH_4), the same basic set of photochemical reaction schemes are expected to dominate in the upper atmospheres of the two planets. Indeed, hydrocarbon photochemical products such as CH_3, C_2H_2, C_2H_4, C_2H_6, CH_3C_2H, C_4H_2 and C_6H_6 have been observed in the stratospheres of both planets [Moses *et al.*, 2001]. An example of the the photochemistry in Saturn's atmosphere is given in Fig. 6.9.

6.2.4   *Greenhouse effect*

As we have mentioned in the first chapter, the equilibrium temperature for Earth is 255 K. The mean surface temperature, however, is 289 K, and that is explained by the natural greenhouse effect. When speaking about planetary habitability or habitability on exomoons, such an effect must therefore be taken into account.

The strength of the greenhouse effect (see Fig. 6.10) depends mainly on:

- inventory of greenhouse gases,
- the albedo effect of clouds,
- the amount of liquid surface water,
- the spectral energy distribution of the host star.

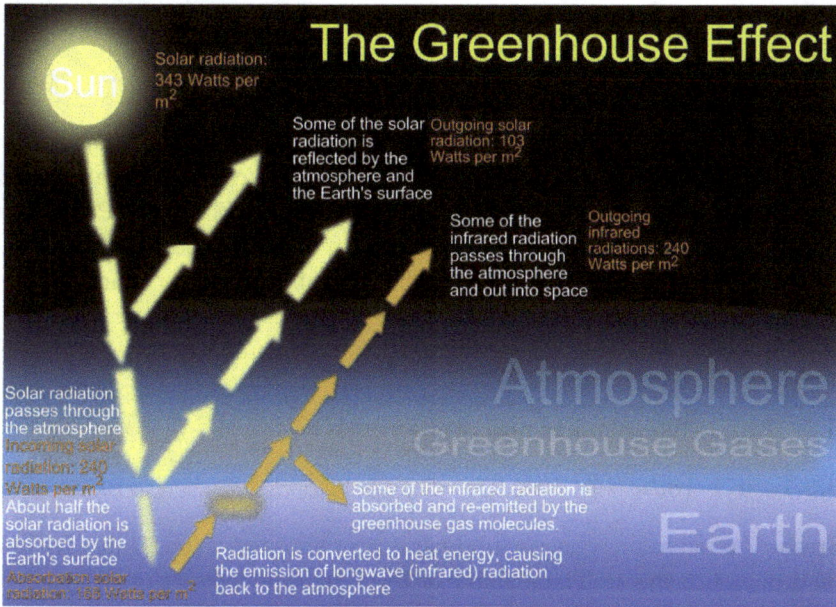

Fig. 6.10   This diagram shows how the greenhouse effect works. Incoming solar radiation to the Earth equals 341 W/m^2 [Trenberth *et al.*, 2009]. Some of the solar radiation is reflected back from the Earth by clouds, the atmosphere and the Earth's surface (102 W/m^2). Some of the solar radiation passes through the atmosphere. About half of the solar radiation is absorbed by the Earth's surface (161 W/m^2). Solar radiation is converted to heat energy, causing the emission of longwave (infrared) radiation back to the atmosphere (396 W/m^2). Some of the infrared radiation is absorbed and re-emitted by heat-trapping "greenhouse" gases in the atmosphere. Outgoing infrared radiation from the Earth equals 239 W/m^2. Credit: Zoo Fari, CC BY-SA3.0.

It can be demonstrated that as the globally absorbed irradiation on a water-rich object increases, the atmosphere gets enriched by water vapor until it becomes opaque. For the Earth this limit is about 300 W/m^2. If the global flux exceeds this limit, a runaway greenhouse (moist greenhouse effect) will result. In this case, water vapor rises from the troposphere to the stratosphere where photodissociation by stellar UV radiation occurs and the light hydrogen atoms or molecules escape into space. The critical flux for runaway greenhouse was discussed by Pierrehumbert [2010]:

$$F_{\mathrm{RG}} = \sigma \left(\frac{l}{R \ln \left(P' / \sqrt{\frac{2}{P_0} og_s(M_s, R_s) k_0} \right)} \right)^4 , \qquad (6.1)$$

with

$$P' = P_{\text{ref}} \exp\left(-\frac{l}{RT_{\text{ref}}}\right), \tag{6.2}$$

where $P_{\text{ref}} = 610.616\,\text{Pa}$, l is the latent heat capacity of water, R the universal gas constant, $T_{\text{ref}} = 273.13\,\text{K}$, $o = 0.7344$ comes from radiative transfer simulations, σ is the Stefan–Boltzmann constant, $P_0 = 104\,\text{Pa}$ is the pressure at which the absorption line strengths of water vapor are evaluated, $g_s = GM_s/R_s^2$ is the gravitational acceleration at the satellite's surface, and $k_0 = 0.055$ is the gray absorption coefficient at standard temperature and pressure. Note that this critical flux does not depend on composition!

6.3   Earth-like planets and stellar activity

6.3.1   *UV radiation and life*

Stellar UV radiation is an essential parameter limiting the possibility for life on a planet or moon. The amount of UV radiation depends on the stellar spectrum and its activity level. We have relatively good information for the Sun concerning its UV output variation during energetic events, like flares.

Let us first consider the UV variation of solar UV radiation during a solar cycle. This was studied by Tourpali *et al.* [2003]. The stratosphere–troposphere system shows a response to a realistic solar cycle enhancement of UV radiation. Stratospheric ozone increases and the stratospheric temperature also increases, changing the stratospheric zonal wind pattern. The authors conclude that the 11-year solar cycle effect on global mean temperature is negligible, but simulated responses of sea level pressure do suggest that regional effects are probably significant, e.g. by affecting the North Atlantic Oscillation.

To study the biological impact of the UV flux, the intensity in the UV wavelength range (λ_1, λ_2) has to be calculated:

$$I_{\text{UV}} = \int_0^T \int_{\lambda_1}^{\lambda_2} I(\lambda)d\lambda dt, \tag{6.3}$$

T means the total time of duration of UV radiation.

As is common when measuring radiation (consider the difference between Sv and gray units), we have to take into account the DNA action spectrum here: the response of biological systems varies with λ. The DNA damage is higher for UVC and decreases considerably for UVB (see also Fig. 6.11).

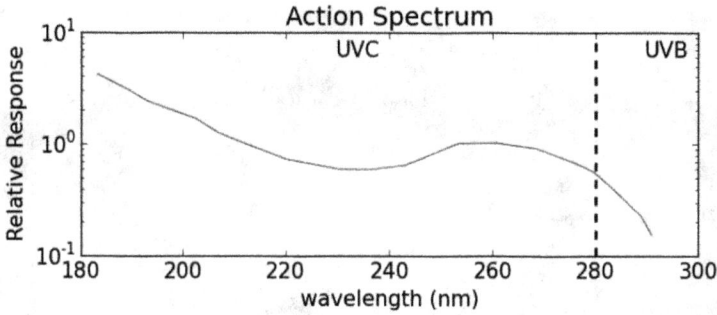

Fig. 6.11   DNA action spectrum for the wavelength range of 182–292 nm. The DNA damage is higher for UVC and decreases considerably for UVB.

Examples can be found in Rugheimer *et al.* [2015]. The biologically effective irradiance can be written as

$$E_{\text{eff}} = \int_{\lambda_1}^{\lambda_2} F_{\text{inc}}(\lambda) S(\lambda) d\lambda. \tag{6.4}$$

F_{inc} is the incident UV flux with the superflare contribution arriving the the planet's surface, S is the action spectra and λ is the MUV wavelengths (200–300 nm). The MUV contribution can be estimated using the Sun as a proxy. For example, the X17 class flare with a total energy of 4×10^{32} erg increased the solar MUV flux by 12% [Woods *et al.*, 2004]. For the 1.8×10^{35} erg superflare, an increase by 5,400% can be estimated.

The UV flux arriving on the surface of a planet is attenuated by the presence of an atmosphere. For Earth, the UV surface irradiance F_{inc} was calculated by Estrela and Valio [2017] for different scenarios:

- an Archean atmosphere with 80% N_2 and 20% CO_2,
- a present-day atmosphere with 80% N_2 and 20% O_2 without ozone,
- a present-day atmosphere with 80% N_2 and 20% O_2 with ozone.

Due to its fainter luminosity, the non-flaring solar irradiation was 75% between 3.5–4 Gyr ago.

6.3.2   *Solar activity and habitability on Mars*

As we discussed in the previous chapters, Mars was most likely a habitable planet during its early history. The history of water on Mars is sketched in Fig. 6.12. Evidence of water in the past and water in the form of ice

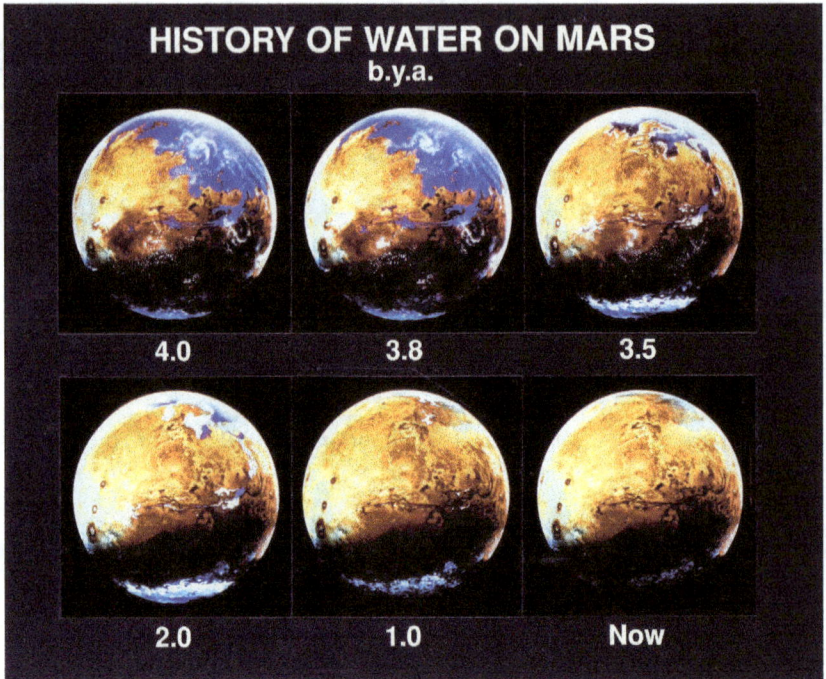

Fig. 6.12 History of water on Mars. Numbers represent how many billions of years ago. Credit: NASA.

present in the Martian soil are discussed in Arvidson [2016]. We give some examples from different space missions:

- Viking Landers found that soils have basaltic compositions, containing minor amounts of salts and one or more strong oxidants.
- Pathfinder with its rover confirmed flood-deposited boulders in Valles Marineris.
- Spirit found evidence for hydrothermal deposits surrounding the Home Plate volcanoclastic feature.
- Opportunity discovered that sandstones originated as playa muds that were subsequently reworked by wind and rising groundwater.

The ancient Martian atmosphere must have been much denser in order to keep liquid water on its surface. Since we have defined a planet as habitable when there is liquid water on its surface, Mars belonged to the class of habitable planets. Why did Mars transform from a habitable planet to an almost inhabitable world?

Studies of the upper planetary atmospheric energetics, which are mainly influenced by solar extreme UV radiation (EUV) were made with the IUVS instrument onboard the MAVEN spacecraft (Mars Atmosphere and Volatile Evolution, launched 2013). A significant drop of 100 K was found in the Martian thermospheric temperature. This can be explained by two effects: (i) drop of solar activity, (ii) increase of the heliocentric distance of Mars. IUVS also observed the response of the Martian thermosphere to the short-term 27-day solar flux variation due to solar rotation. Moreover, the effect of two solar flare events (October 19, 2014 and March 24, 2015) on Martian dayglow observations was reported. During these events, IUVS observed about a 25% increase in observed brightness of major ultraviolet dayglow emissions below 120 km, where most of the high-energy photons (<10 nm) deposit their energy [Jain *et al.*, 2016].

Thus, MAVEN showed that the deterioration of Mars' atmosphere increases significantly during solar storms. Today, there is no liquid water on the Martian surface. Mars loses water into its thin atmosphere through evaporation, and solar UV radiation can split the water molecules into their components, hydrogen and oxygen. The hydrogen, as the lightest element, then tends to rise far up into the highest levels of the Martian atmosphere, where several processes can strip it away into space, to be forever lost to the planet. This loss was thought to proceed at a fairly constant rate, but MAVEN's observations of Mars's atmospheric hydrogen through a full Martian year (almost two Earth years) show that the escape rate is highest when Mars's orbit brings it closest to the Sun, and only one-tenth as great when it is at its farthest point from the Sun (aphelion). That loss of atmosphere to space likely played a key role in Mars' gradual shift from its carbon dioxide-dominated atmosphere, which had kept Mars relatively warm and allowed the planet to support liquid surface water to the cold, arid planet seen today. This shift took place between about 4.2–3.7 billion years ago. In Fig. 6.13, the processes of atmospheric mass loss are shown. An example of the MAVEN measurements for the loss of Martian atmosphere is shown in Fig. 6.14.

The history of the outgassing and escape of the Martian atmosphere and water inventory was reviewed in Lammer *et al.* [2013]. They divided the evolution and escape of the Martian atmosphere and the planet's water inventory into an early and late evolutionary epoch.

- The first epoch started from the planet's origin and lasted 500 Myr. Due to the high EUV flux of the young Sun and Mars' low gravity, it

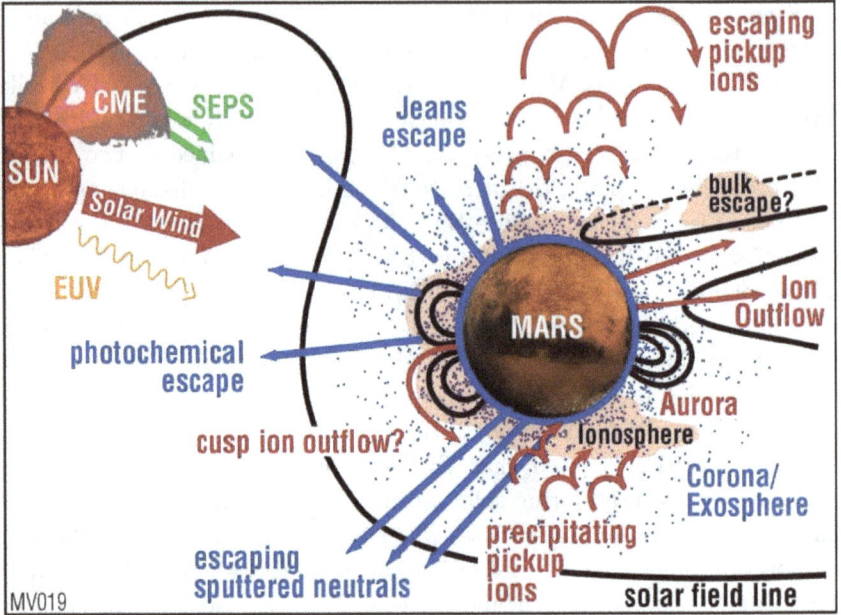

Fig. 6.13   MAVEN measured the process affecting the remaining atmosphere on Mars. These include incoming solar energetic particles (SEPs), escape on a molecule-by-molecule basis (Jeans escape), the effect of coronal mass ejections (CMEs) and extreme solar ultraviolet radiation (EUVs). Credit: The Lunar and Planetary Institute and LASP.

Fig. 6.14   Three views of an escaping atmosphere, obtained by MAVEN's Imaging Ultraviolet Spectrograph, are shown here. By observing all of the products of water and carbon dioxide breakdown, MAVEN's remote sensing team can characterize the processes that drive atmospheric loss on Mars. These processes may have transformed the planet from an early Earth-like climate to the cold and dry climate of today. Credit: NASA/MAVEN.

was accompanied by a hydrodynamic blow-off of hydrogen and strong thermal escape rates of dragged heavier species such as O and C atoms. Through these processes, the main part of the protoatmosphere was lost.

- Impact-related volatiles and mantle outgassing may have resulted in the accumulation of a secondary CO_2 atmosphere of a few tens to a few hundred mbar around 4–4.3 Gyr ago.

6.3.3 *The case of Kepler-96b*

On early Earth there was no free oxygen in its atmosphere because there was no photosynthesis. It is assumed that a considerable amount of oxygen has accumulated in the Earth's atmosphere at an age of about 2.3 Gyr. This increase of oxygen occurred because of the photosynthetic activity of microorganisms living under the sea.

The exoplanet Kepler-96b is of special interest in this context. Its radius is about $0.2\,R_J$ and its mass about $0.02\,M_J$. Its orbital period is only about 16 days, the distance from the host star about 0.13 AU. The host star radiation at the atmospheric boundary of the planet is about 85,000 W/m² which is about 60 times the amount we receive on Earth. The host is one time more massive and one time bigger compared with the Sun. Its temperature is about the same as the Sun.

The star Kepler-96 was investigated because it shows superflares. The biggest flares found released an energy of about 1.8×10^{35} erg.

In the paper of Estrela and Valio [2017] more than 80 transits were analyzed, some showing clear signatures of superflares. In one case such a superflare led to a stellar flux increase of 4%.

If t_0 denotes the peak time of the flare, a Gaussian profile can be used to describe it:

$$I_{\text{flare}} = A \exp \left[\frac{(t - t_0)^2}{2\sigma_t^2} \right], \tag{6.5}$$

where A denotes the amplitude of the flare and σ_t is the duration of the flare. The total energy release from the flare becomes

$$E_{\text{flare}} = AL_* \int_{-\infty}^{+\infty} \exp((t - t_0)^2 / 2\sigma_t^2) dt. \tag{6.6}$$

For example, for the superflare that increased the UV flux by 5,400%, the values given in Table 6.3 were found (adapted from Estrela and Valio [2017]).

Table 6.3. Biological effective irradiance from Kepler-96, E_{eff} (J/m^2).

Object	No atm.	Archean atm.	Present atm. without O$_3$	Present atm. with O$_3$
Kepler-96b	3.2×10^8	3.4×10^7	5×10^6	178
Planet at 1 AU	1.5×10^5	1.6×10^4	2401	0.0084

It is claimed that only extremophile life could survive on the surface of Kepler-96b if there was an ozone layer present on the planet atmosphere. However, under the Archean conditions, lifeforms such as *E. Coli* could widthstand the effects of the strongest superflare on this planet if they were at a depth of 48 m below the ocean surface. For a hypothetical Earth, depending on the size of the superflare and the resistance of the microorganism, life could be sustained on the surface even in an Archean atmosphere or at an ocean depth of 8–20 m.

6.3.4 *Biosignatures*

Any small amount of molecular oxygen in an Earth-like planet's atmosphere produced by photolysis of water vapor is consumed by the oxidation of surface rocks and volcanic gases. Thus, if oxygen and liquid water are simultaneously observed in a spectrum, there must be some additional source producing the oxygen. The most likely source would be oxygenic photosynthesis. If ozone and liquid water are seen in a spectrum, it would be a very strong biosignature. The formation of ozone (O$_3$) requires the presence of oxygen in the planet's atmosphere, since UV radiation dissociates molecular oxygen, which then recombines to form ozone. Ozone has a spectral signature in the infrared part of the spectrum, making it easier to detect than oxygen (which is detected at visible wavelengths). The synthetic spectrum of an Earth-like planet is shown in Fig. 6.15.

The effect of life on Earth is shown by a comparison of the three terrestrial planets Venus, Earth and Mars (Fig. 6.16).

Spectral signatures of photosynthesis were discussed in the paper of Kiang *et al.* [2007]. Earth-like planets around observed F2V and K2V, modeled M1V and M5V stars, and around the active M4.5V star AD Leo were investigated by two scenarios: (i) Earth's atmospheric composition (ii) as well as very low O$_2$ content in case anoxygenic photosynthesis dominates. Photon flux densities on the surface of the planet and under water were calculated, and bands of available photosynthetically relevant radiation were identified. Photosynthetic pigments on planets around F2V stars may

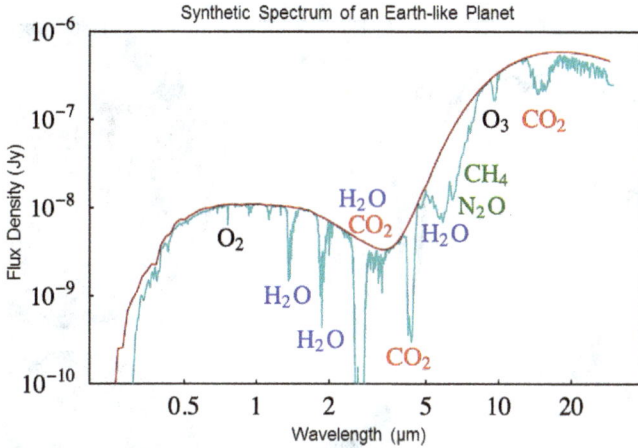

Fig. 6.15 Synthetic spectrum of an Earth-like planet with biosignatures (H_2O, CO_2). Courtesy: H. Rauer *et al.* (2011). Potential biosignatures in super-Earth atmospheres, *Astronomy & Astrophysics*, 16.

peak in absorbance in the blue, K2V in the red–orange, and M stars in the near-infrared, in bands at 0.93–1.1 μm, 1.1–1.4 μm, 1.5–1.8 μm, and 1.8–2.5 μm. However, underwater organisms are restricted to wavelengths shorter than 1.4 μm. Under water, organisms would still be able to survive ultraviolet flares from young M stars and acquire adequate light for growth.

Spectropolarimetry can be of great importance for detecting life in the universe because photosynthetic life can be detected on other planets in visible polarized spectra with high sensitivity [Berdyugina, 2016]. In this study, it is shown that the polarization at a given wavelength in the planetary atmosphere arises from anisotropic irradiation and unequal illumination coming from the top and bottom.

- Near the bottom of a planetary atmosphere, the thermal radiation of the planet dominates; anisotropy and polarization are small.
- near the top of a planetary atmosphere: stellar irradiation dominates.

The anisotropy causes polarization. Hotter stars hosting close planets produce larger polarization in the blue. The polarization near the limb of the planet's disk is very sensitive to stellar radiation; for a larger stellar flux, it increases. Polarization measurements could even be used to detect stellar variability (e.g. spots) because the signal strongly depends on the incoming flux. Photosynthetic organisms absorbing visible stellar radiation with the help of various biopigments demonstrate a high degree of linear polarization associated with absorption bands in the spectra.

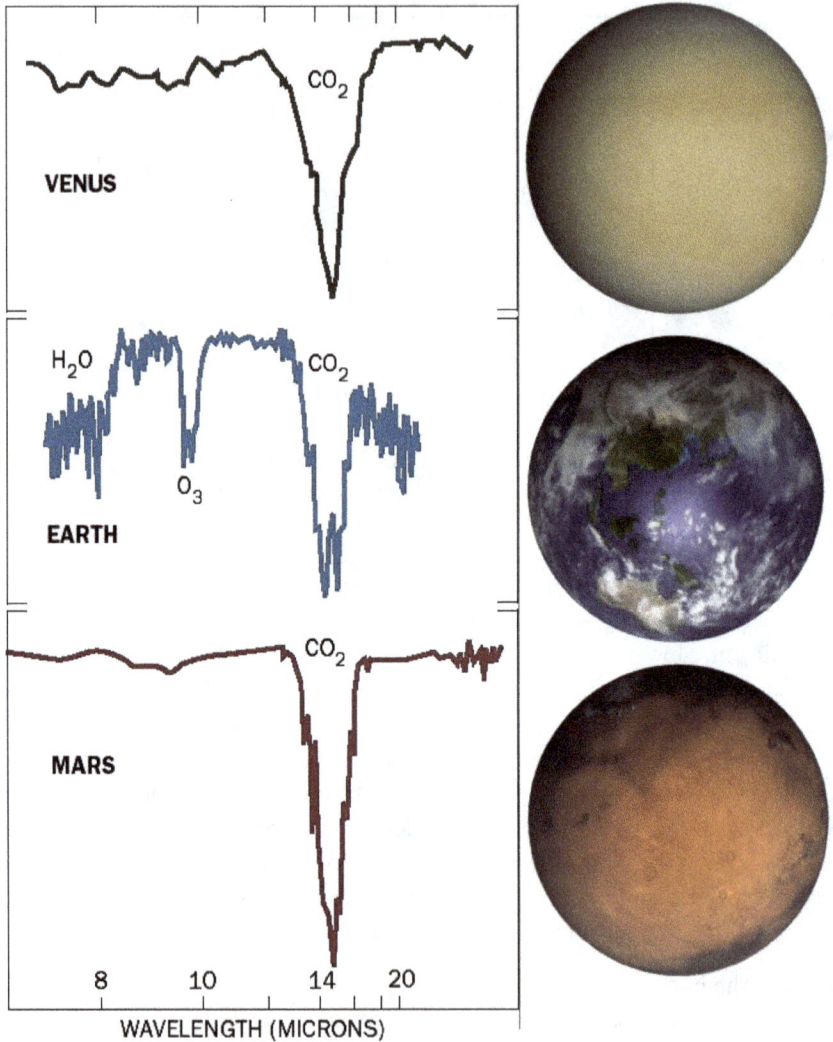

Fig. 6.16 Synthetic spectrum of an Earth-like planet with biosignatures (H_2O, CO_2, O_3). Credit: NMSU Astronomy.

6.4 Post-main sequence stars and planets

In this section we give an overview on stars that have evolved away from the main sequence and their planets. How does the transition from a relatively quiet main sequence star to a red giant and subsequent transition to a white dwarf influence on a planetary system orbiting such stars? What do we know from observations?

6.4.1 *Post-main sequence evolution and dynamical evolution of planets*

As we have outlined in the chapter about stellar evolution, a low-mass star evolves into a red giant and finally into a white dwarf.

What happens to planets orbiting close to such stars during their transition from a main sequence star to a red giant?

Generally, it can be summarized that as a star evolves, planetary orbits change over time owing to:

- tidal interactions,
- stellar mass loss,
- friction and gravitational drag forces,
- mass accretion, and evaporation on/by the planet.

The stellar mass loss can be estimated from the empirical relation [Reimers, 1975]

$$\dot{M} = 4 \times 10^{-13} \eta L_* R_* / M_* \qquad M_\odot \text{yr}^{-1}, \tag{6.7}$$

where $\eta = 0.5$ [Maeder and Meynet, 2012] and L_*, M_* are the luminosity and the mass of a star. The mass loss drives planetary orbit evolution. The evolution of the semi-major axis a of a planetary orbit (under the assumption of a circular orbit, $e = 0$ and aligned with the equator of the star) can be given by

$$\left(\frac{\dot{a}}{a}\right) = -\frac{\dot{M}_* + \dot{M}_{\text{pl}}}{M_* + M_{\text{pl}}} - \frac{2}{M_{\text{pl}} v_{\text{pl}}} [F_{\text{fri}} + F_{\text{grav}}] - \left(\frac{\dot{a}}{a}\right)_t. \tag{6.8}$$

In this equation $M_{\text{pl}}, \dot{M}_{\text{pl}}$ are the planetary mass and the rate of change in the planetary mass, v_{pl} is the velocity of the planet, $F_{\text{fri}}, F_{\text{grav}}$ are the frictional and gravitational drag forces, and the last term $(\dot{a}/a)_t$ takes into account the effects of the tidal forces. The tidal term is responsible for the exchange of angular momentum between the planet and the star. In these terms the rotation of the star is involved. Tidal dissipation can be very efficient in the stellar envelope when a convective envelope evolves. We have already discussed that the depth of a convective envelope, i.e. the surface convection zone, becomes larger for cooler stars. The transfer of angular momentum from the planetary orbit to the star or the inverse depends on whether the orbital angular velocity of the planet is smaller or larger than the angular rotation of the star. In Nordhaus and Spiegel [2013] the following term is given

$$\left(\frac{\dot{a}}{a}\right) = \frac{f}{\tau}\frac{M_{\mathrm{env}}}{M_*}q(1+q)\left(\frac{R_*}{a}\right)^8\left(\frac{\Omega_*}{\omega_{\mathrm{pl}}}-1\right), \tag{6.9}$$

with M_{env} the mass of the convective envelope, $q = M_{\mathrm{pl}}/M_*$, ω_{pl} the orbital angular velocity of the planet, and Ω_* the angular velocity of the stellar surface. f is given by the ratio of the orbit half period $P/2$ and the convective eddy turnover time τ. This parameter is given by

$$\tau = \left[\frac{M_{\mathrm{env}}(R - R_{\mathrm{env}})^2}{3L_*}\right]^{1/3}. \tag{6.10}$$

This formula was given by Rasio *et al.* [1996].

The evolution of a two-body system wherein the orbit adjusts due to structural changes in the primary component, dissipation of orbital energy via tides, and mass-loss during the giant phases and changes in the primary's spin was studied by Nordhaus and Spiegel [2013]. When an Earth-like planet becomes engulfed, it cannot survive a common envelope phase. A first-generation planet in a white dwarf habitable zone requires scattering from a several AU orbit to a high eccentricity orbit (periastron near 1 $R_\odot$ from which it is damped into a circular orbit via tidal friction). However, this object will most probably be not habitable. In Fig. 6.17 the escape distances of exoplanets for stars with different masses are given. In Fig. 6.18 the eventual fates of known exoplanets are shown.

6.4.2 *Solar post-main sequence evolution and habitability on Earth*

The Sun's post-main sequence evolution can be described by the two steps strongly affecting the habitability in our planetary system. They are as follows:

- On the red giant branch (RGB), the solar diameter increases but does not quite reach the Earth's orbit. However, due to the strongly increased solar surface as a result of its expansion, its luminosity also increases by several orders of magnitudes. The Earth would no longer be a habitable planet.
- On the asymptotic giant branch (AGB), the Sun would engulf Earth.

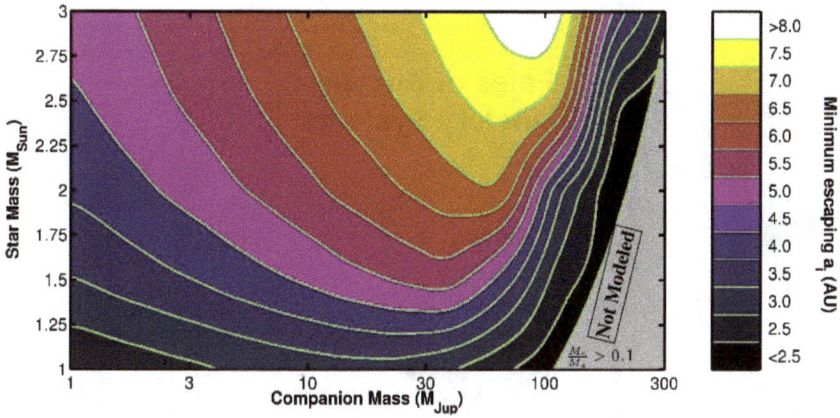

Fig. 6.17   Escape distances for stars and planets of different masses. The green curves are lines of equal escape distance. The color code gives the escape distance in AU. Escape distances are smallest for large planets and small stars. Credit: Nordhaus and Spiegel.

Such a simple scenario, however, does not take into account the mass loss that becomes very strong during these phases of stellar evolution. If the star loses mass, the orbits of the planets will become larger because of conservation of angular momentum. Calculations show that the Sun will lose about 40% of its mass (i.e. it will have only 0.59 $M_\odot$) when it expands to 0.99 AU. As a consequence, the Earth's orbit will have $a = 1.69$ AU. There, the Earth will survive as a planet.

The most direct determination of the compositions of extra-Solar planets, asteroids and comets is, in fact, made by an analysis of the elemental abundances of the remnants of these bodies accreted into the atmospheres of white dwarfs. To understand how the accreted bodies relate to the source populations in the planetary system, and to model their dynamical delivery to the white dwarf, it is necessary to understand the effects of stellar evolution on bodies' orbits. On the RGB and AGB prior to becoming a white dwarf, stars expand to a large size (>1 AU) and are easily deformed by orbiting planets, leading to tidal energy dissipation and orbital decay. They also lose half or more of their mass, causing the expansion of bodies' orbits. This mass loss increases the planet–star mass ratio, so planetary systems orbiting white dwarfs can be much less stable than those orbiting their main-sequence progenitors. Finally, small bodies in the system experience strong non-gravitational forces during the RGB and AGB: aerodynamic drag from the mass shed by the star, and strong

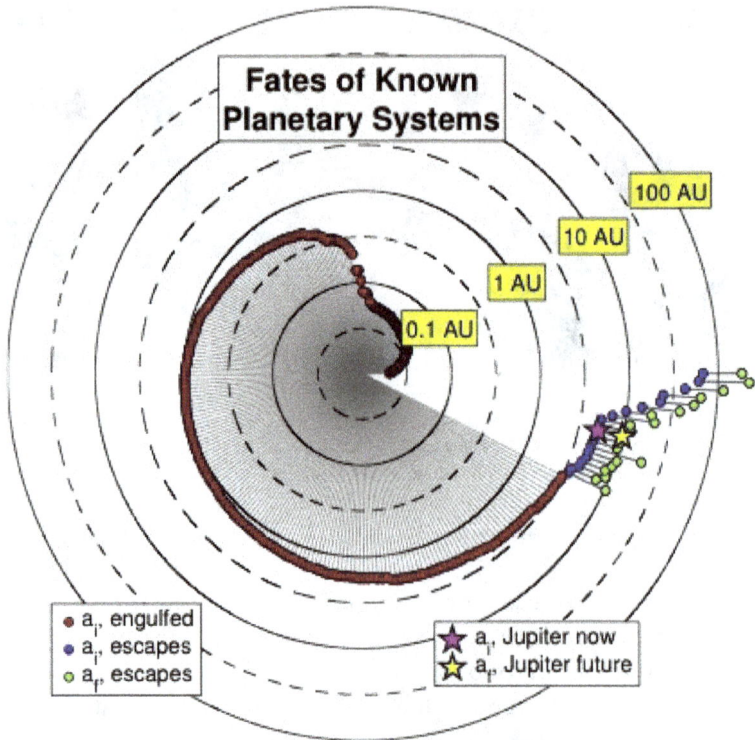

Fig. 6.18 The eventual fates of the planets. The red planets will be engulfed by the red giant and destroyed. The blue planets represent the initial positions of the surviving, non-engulfed planets. These planets will move outward in their orbits due to mass loss; their final positions are in green. The yellow star represents Jupiter for scale. Credit: Nordhaus and Spiegel.

radiation forces as the stellar luminosity reaches several thousand solar luminosities.

Another important parameter that describes stellar activity is the rotation period of a star. How does this parameter change when the Sun evolve into a red giant and beyond?

The orbital angular momentum of a planet around its host star can be expressed as

$$L_p = m_p \sqrt{GMa(1 - e^2)}, \qquad (6.11)$$

where m_P is the planet's mass, M the mass of its host star, a the orbital semi-major axis and e the orbital ellipticity. We have already seen that when the Sun loses mass, the planets are pushed outward.

Several effects come into play:

- Solar wind particles collide with the planet;
- Planet may accrete up to 1% of its mass from this strong solar wind;
- Planetary outgassing and loss of atmosphere. This effect is negligible for Earth because of the low-mass fraction of its atmosphere; however, Earth will lose its atmosphere completely.

For the change of the semi-major axis, we can write

$$\frac{1}{a}\frac{da}{dt} = -\frac{1}{M}\frac{dM}{dt} \tag{6.12}$$

or

$$a = \frac{M_\odot}{M}a_0, \tag{6.13}$$

where M denotes the mass of the Sun at time t, a_0 the semi-major axis of a planet's orbit at present. During the RGB phase, the Sun will expand about 200 times. The tidal force will have two consequences:

- The inner planets move inward;
- Transfer of angular momentum from the inner planets to the solar envelope.

Our Sun will expand enormously and lose substantial mass via a stellar wind during the RGB phase; the rotational period will be prolonged by several orders of magnitude.

In Figs. 6.19–6.21 the evolution of the Sun's mass (present day value = 1), radius (present day value = 1) and rotation period (in years) between the solar ages of 11.0 and 12.6 Gyr are shown.

To give some concrete numbers about the RGB evolution of the Sun:

- The RGB phase is from 11.98 Gyr to 12.52 Gyr.
- The total duration of the RGB phase is 543 Myr.
- The RGB phase starts when $R = 2.46R_\odot$.
- From $R = 2.46R_\odot$ to $R = 20R_\odot$ the duration is 516 Myr. The rotation period is less than 48 yr.
- From $R = 100R_\odot$ to $R = 196R_\odot$ the duration is very short, 2.6 Myr.

It is difficult to predict how much mass the Sun will lose before it reaches the RGB tip. In the work of Guo *et al.* [2017], the Sun is considered as a two-component system comprised of a core and a convective envelope, each

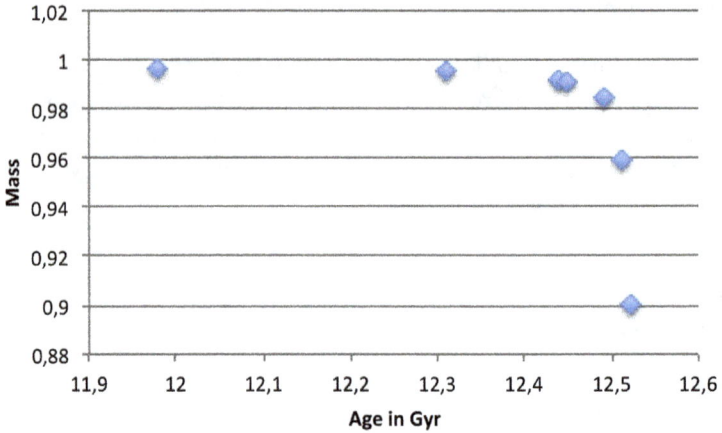

Fig. 6.19   Evolution of the Sun's mass. Adapted from Guo *et al.* [2017].

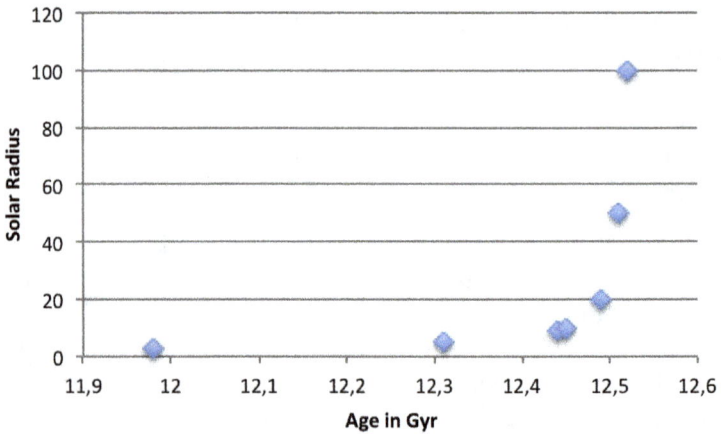

Fig. 6.20   Evolution of the Sun's radius. Adapted from Guo *et al.* [2017].

being allowed to rotate freely. The angular momentum transfer from the inner planets to the solar envelope is taken into consideration. A formula like

$$\dot{M} = 4 \times 10^{-13} \eta L_* R_* / M_* \qquad M_\odot \mathrm{yr}^{-1} \qquad (6.14)$$

was used and the solar envelope's rotational period at the RGB tip varies from 1,792 to 736,934 years, as the Reimers η is changed from 0.00 to 0.75. Recent observations show that the average Reimers η of Sun-like stars is 0.477. Adopting this average value of the Reimers η, the solar envelope's rotational period at the RGB tip will be 24,868 years. Other Sun-like stars,

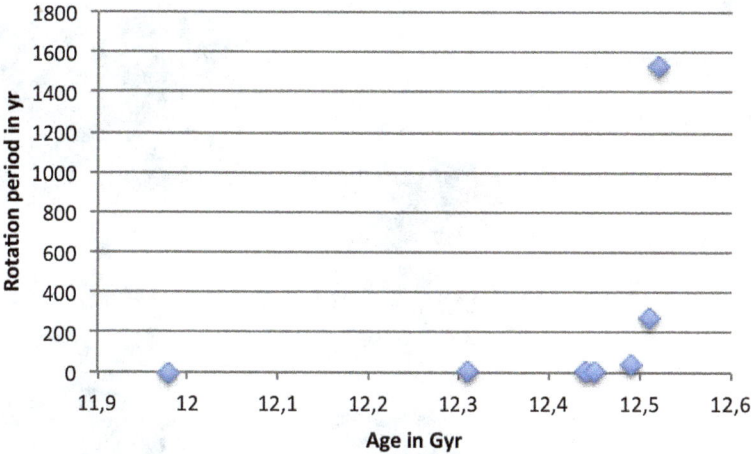

Fig. 6.21 Evolution of the Sun's revolution period (envelope). Adapted from Guo *et al.* [2017].

with different planetary configurations, may prematurely eject mass and lead to planetary nebulae, if they engulf a brown-dwarf companion at the RGB tip. By swallowing a planet with 13 Jupiter masses and a 3-day orbit, a Sun-like star can become a rapidly rotating giant star.

The habitable zone in the solar system, once the Sun has reached its RGB phase, is shown in Fig. 6.22. It is seen that Jupiter and even Saturn will become habitable planets.

6.4.3 *Rapidly rotating giant stars*

It is predicted that the Sun will rotate extremely slowly during its RGB evolution.

Hot Jupiters typically have orbital periods of about 3 days. We can apply Kepler's third law to derive the semi-major axis for such a planet: $a = 0.0407\,\mathrm{AU} = 8.75\,R_\odot$. The angular momentum for a planet with 1 Jupiter mass is $1.7 \times 10^{42}\,\mathrm{kgm^2/s}$; for a planet with 13 Jupiter masses this value becomes $2.2 \times 10^{43}\,\mathrm{kgm^2/s}$. Let us assume that the Sun expands to $8.75\,R_\odot$. It has already lost some of its mass; therefore, the Sun's mass is about $0.9922\,M_\odot$. The envelope's rotational inertia becomes $8.53 \times 10^{46}\,\mathrm{kgm^2}$, and the envelope's angular momentum $1.63 \times 10^{41}\,\mathrm{kgm^2/s}$; the envelope's rotational period becomes 10.4 yr. If the Sun engulfs a planet with 1 Jupiter mass, its envelope's rotational period would become 331.16 days; if it engulfs

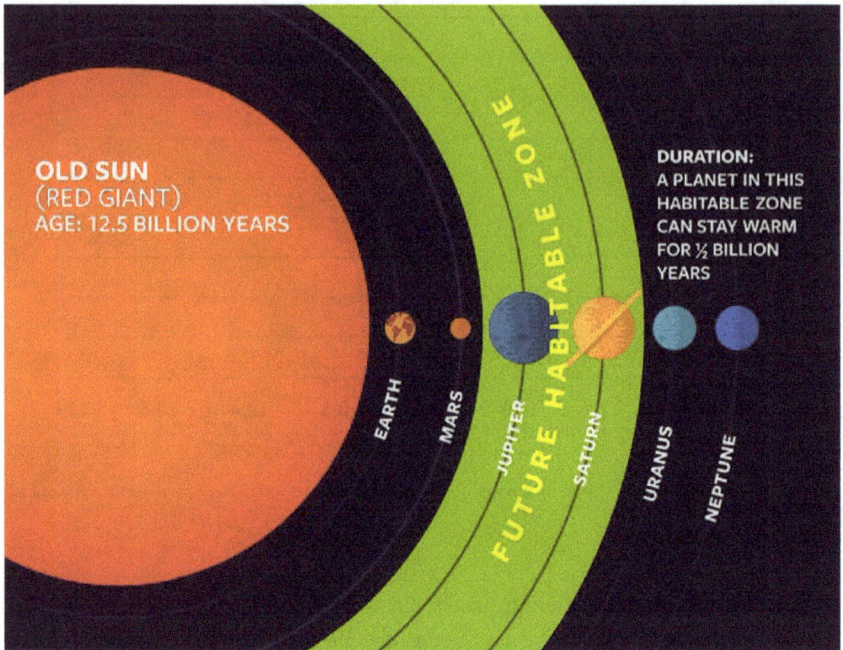

Fig. 6.22 A diagram of where our solar system's new habitable zone will reside after the Sun becomes a red giant. Credit: Cornell University, Wendel.

a planet with 13 Jupiter masses, the envelope's rotational period would be about 27 days!

In Costa *et al.* [2015] seven Sun-like rapidly rotating giant stars, with radii from 6.36 to 11.08 $R_\odot$ and rotational periods from 13.93 to 54.74 days are discussed. Comparing these values with the above consideration about the Sun, we can explain such rotational periods for late-type stars only through the process of engulfing close planets (3-day orbits) with several Jupiter masses. Certainly, by such engulfment, the planets will no longer be habitable.

6.5   Hot Jupiters

6.5.1   *Why study hot Jupiters?*

Since we focus on stellar activity and habitability, the class of hot Jupiters should be excluded from this discussion. However, since hot Jupiters are quite close to their hosting stars, it is assumed that they can influence and trigger the activity of these host stars. The star–planet interaction could be explained by

- magnetic interaction,
- or tidal interaction.

Examples of such cases are the stars μ And, τ Boo or HD 179949 or HD 189733.

Transit observations can be used to probe the atmosphere of hot Jupiters. These observations clearly showed that by photoionization, the atmospheres of these hot Jupiters get heated and magnetized winds are driven from their atmospheres. The estimated mass-loss rates are expected on the order of $10^7 - 10^9$ kg/s. This expanding gas interacts with the stellar wind plasma that propagates with a few hundred km/s.

Using MHD simulations [Matsakos *et al.*, 2015], the interaction of these two gas components can be studied. The planetary gas forms a stream of material that gets compressed and accretes onto the star with a phase lag of 70–90°. In both cases, regardless of whether the stellar wind collides with the planetary outflow or its magnetosphere, the interaction can have a potentially observable signature:

- A bow shock in front of the planet;
- A cometary like structure might be formed, trailing the orbit.

In both cases, an absorption of stellar light should be observable. The observations can be made, for example, in the X-ray and FUV regime.

6.5.2   *Example: HD 189733*

This is an example of a very well-known transiting hot Jupiter. It is a binary system. The host star is estimated to be around 5 Gyr, but its activity was found to be relatively high.

The parameters of the system are given according to Pillitteri *et al.* [2015] in Table 6.4.

The triggering of stellar activity by a closely orbiting planet was studied. Flaring activity, for example, was observed with the XMM-Newton instrument in X-rays. The triggering of such flaring activity due to a planet can be seen from the enhanced flaring activity detected near the eclipses (at phases near 0.52 and 0.65). With the HST instrument, observations were made in the FUV. An increase in the Lyman α profile was observed, as well as in the Si III, Si IV, C II and other lines.

Table 6.4. Parameters for the system HD 189733. Adapted from Pillitteri *et al.* [2015].

	HD 189733A	HD 189733b	HD 189733B
Type	K 1.5 V	Planet	M4V
Mass	0.81 $M_\odot$/1.15 M_J	0.2 $M_\odot$	—
Radius	0.76 $R_\odot$	1.26 R_J	—
Orbital period	—	2.219 d	3200 yr
Mean orbital radius	—	0.003 AU	216 AU
Age	$\geq$5 Gyr	—	$\geq$5 Gyr

6.5.3   *HD 209458b*

In the paper by Charbonneau *et al.* [2000], high-precision and high-cadence photometric measurements of the star HD 209458, which is known from radial velocity measurements to have a planetary-mass companion in a close orbit, were presented. Thus, HD 209458b became the first exoplanet to be detected by the transition method (i.e. decrease in the brightness of the host star due to the transit of the hot Jupiter planet).

In del Burgo and Allende Prieto [2017] the parameters were summarized: The stellar diameter can be obtained from interferometric measurements. An improvement could be made by including specific model atmospheres (Kurucz model), and the value obtained for the angular diameter θ was: $\theta = 0.2254 \pm 0.0017$ mas.

This angular diameter represents an improvement in precision more than four times greater than available interferometric determinations. The stellar radius is $R_* = 1.20 \pm 0.05\ R_\odot$. The radius of the exoplanet HD 209458 b is $R_{\mathrm{p}} = 1.41 \pm 0.06\ R_J$ which is derived from the corresponding transit depth in the lightcurve and the precise stellar radius. The effective temperature of the star is $T_{\mathrm{eff}} = 6071 \pm 20$ K. For the planet HD 209458 b, the corresponding parameters are given in Table 6.5.

In Fig. 6.23 an artist's impression of the evaporating atmosphere of this planet is shown.

Table 6.5.   Parameters for HD 209458b.

Parameter	Value (uncertainty)
R_P (R_J)	1.41 ± 0.06
M_P (M_J)	0.74 ± 0.06
ρ_P (g/cm^3)	0.32 ± 0.05
a (AU)	0.0490 ± 0.0020

Fig. 6.23 An example of an evaporating hot Jupiter: HD 209458b. Credit: ESA/HST, artist's impression, CC-BY 4.0.

6.5.4 *Close inner planets*

First we define the Roche lobe. Let us consider a host star and a planet orbiting it. The Roche lobe is the region around a star within which orbiting material is gravitationally bounded to the star. It is an elongated teardrop-shaped region; the apex of the teardrop points toward the planet, which is identical to the Lagrangian point L_1 of the system.

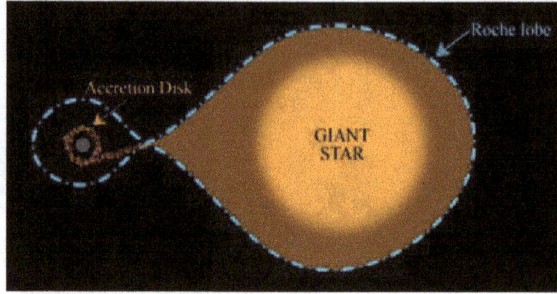

Fig. 6.24   The Roche lobe. Credit: Swinburne University of Technology.

If r_1 is the radius of the Roche lobe (Fig. 6.24), A the orbital separation of the system, and M_1, M_2 are the masses of the host star and the planet, then

$$\frac{r_1}{A} = 0.38 + 0.2 \log \frac{M_1}{M_2}, \qquad 0.3 < \frac{M_1}{M_2} < 20, \qquad (6.15)$$

$$\frac{r_1}{A} = 0.46224 \left(\frac{M_1}{M_1 + M_2}\right)^{1/3}, \qquad \frac{M_1}{M_2} < 0.8. \qquad (6.16)$$

A more adequate formula for the case discussed here (planet and host star) is

$$\frac{r_1}{A} = \frac{0.49q^3}{0.6q^{2/3} + \ln(1 + q^{1/3})}, \qquad q = M_1/M_2. \qquad (6.17)$$

There are several mass loss effects on short period exoplanets that are close to their parent stars.

- Mass loss from their atmospheres (in the case of hot Jupiters their envelope) by the intense X-ray and extreme UV (XUV) radiation from the parent star.
- Roche lobe overflow due to the evolution of their orbits.

These two effects will change the close planets in the following ways:

- Hot Jupiters could retain their envelopes.
- Super-Earths will lose their envelopes completely.
- Hot Jupiters with small cores $\sim 10\, M_\oplus$ may evolve into sub-Jupiter deserts.

As a result of these processes, it may be concluded that populations of closely orbiting exoplanets are formed by capturing planets at/inside the

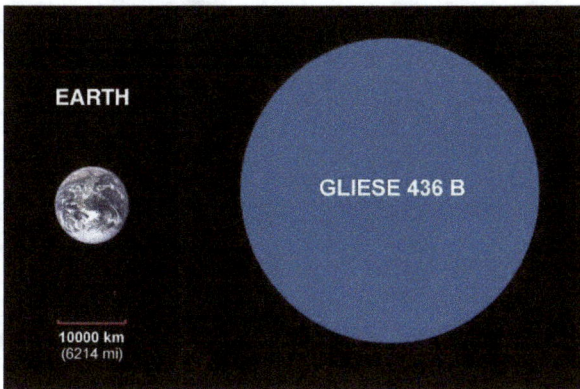

Fig. 6.25   Comparison of Earth and GJ 436b.

inner edges of protoplanetary disks and subsequent evaporation of sub-Jupiters [Kurokawa and Nakamoto, 2014]. A correlation exists between planetary orbit and planetary mass. Jupiter-sized planets form at distances ≥ 1 AU. Therefore, no Jupiter-sized planets should occur within 1 AU. Jupiter-sized planets form at large distances and migrate inward. The detection of the so-called sub-Jupiter desert at ≤ 0.04 AU can only be explained by the abovementioned evaporation processes.

The exoplanet GJ 436 b (Fig. 6.25) belongs to the class of warm Neptunes. A giant exosphere was also detected in this case. During transits, absorption features in the hydrogen Lyman α were detected. Its diameter is about 54,000 km and its mass is 0.07 M_J [Ehrenreich et al., 2015]. The ultraviolet transits repeatedly start about 2 hours before and end more than 3 hours after the approximately 1 hour optical transit, which is substantially different from a previous claim (based on an inaccurate ephemeris). Thus, this object appears to be a huge comet orbiting a red dwarf star (Fig. 6.26). The semi-major axis is 0.03 AU; the eccentricity is 0.15. The distance of GJ 436 is about 10 pc.

6.5.5   *Exoplanet populations*

There seem to be several populations of exoplanets that can be seen in a diagram when plotting e.g. orbital period versus size (Fig. 6.27). These objects include:

- Hot Jupiters: gas giants close to the parent star.
- Cold gas giants: gas giants at larger distances from the parent star.

Fig. 6.26   GJ 436b, a huge comet around a red dwarf star. Credit: NASA, Artist's impression.

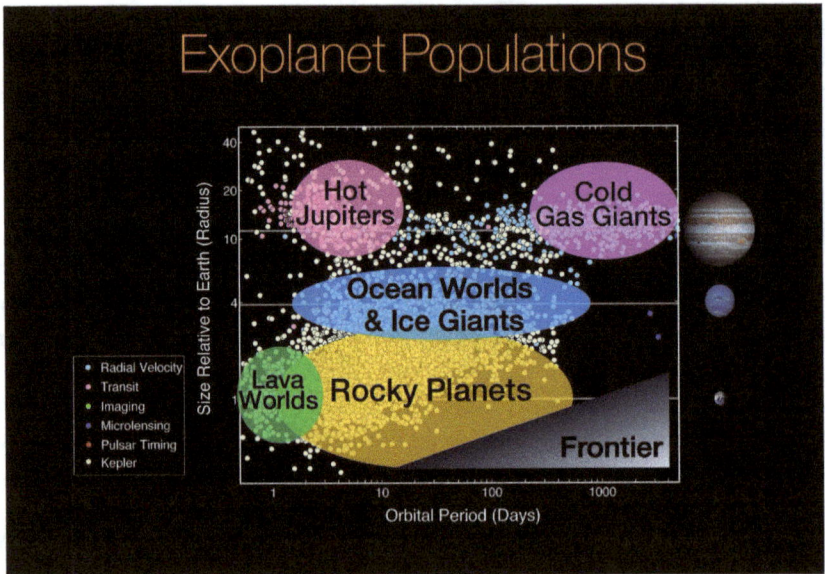

Fig. 6.27   The population of exoplanets detected by the Kepler mission (yellow dots) compared to those detected by other surveys using various methods: radial velocity (light blue dots), transit (pink dots), imaging (green dots), microlensing (dark blue dots), and pulsar timing (red dots). For reference, the horizontal lines mark the sizes of Jupiter, Neptune and Earth, all of which are displayed on the right side of the diagram. The colored ovals denote different types of planets: hot Jupiters (pink), cold gas giants (purple), ocean worlds and ice giants (blue), rocky planets (yellow), and lava worlds (green). The shaded gray triangle at the lower right marks the exoplanet frontier that will be explored by future exoplanet surveys. Kepler has discovered a remarkable quantity of exoplanets and significantly advanced the edge of the frontier. Credit: NASA/Ames Research Center/Natalie Batalha/Wendy Stenzel.

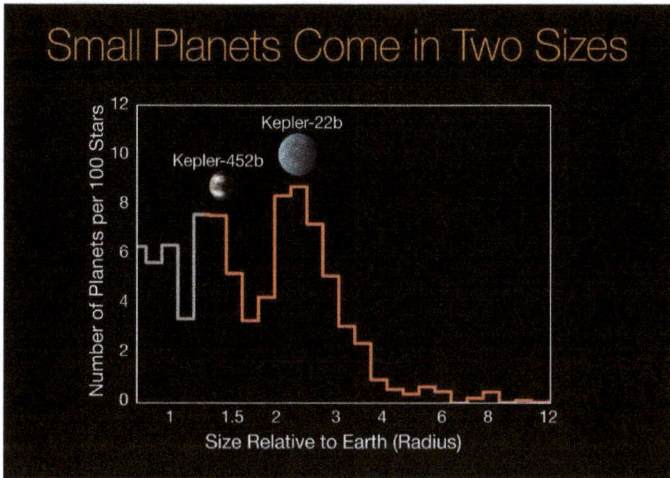

Fig. 6.28 Researchers using data from the W. M. Keck Observatory and NASA's Kepler mission have discovered a gap in the distribution of planet sizes, indicating that most planets discovered by Kepler so far fall into two distinct size classes: the rocky Earth-size and super-Earth-size (similar to Kepler-452b), and the mini-Neptune-size (similar to Kepler-22b). This histogram shows the number of planets per 100 stars as a function of planet size relative to Earth. Credit: NASA/Ames Research Center/CalTech/University of Hawaii/B.J. Fulton.

- Rocky planets.
- Ocean worlds.
- Ice giants.
- Lava worlds: rocky planets very close to parent star.

In this diagram the different detection methods used are also marked. Most objects were found with Kepler.

In Fig. 6.28 it is shown that small exoplanets exist in two populations: Earth/super-Earth populations and mini Neptunians.

Can a significant mass loss by such evaporation processes cause a migration of planets or even a transition of classes i.e. from hot Jupiters to hot Neptunes? These two classes of exoplanets form distinct populations. Mass evaporation can cause an increase in orbit period and can explain therefore a transition from a hot Jupiter object to a hot Neptunian exoplanet.

Let us assume a planet with mass $m(t)$ is orbiting a star with mass m_* (see the paper of Boué *et al.* [2012]). $\mathbf{r}$ is the radius vector, $\mathbf{v} = \dot{\mathbf{r}}$ the orbital velocity. The equations of motion become

$$m\frac{d\mathbf{v}_p}{dt} = -Gm_*m\frac{\mathbf{r}}{r^3} + \dot{m}\mathbf{v}, \tag{6.18}$$

$$m_*\frac{d\mathbf{v}_*}{dt} = Gm_*m\frac{\mathbf{r}}{r^3}. \tag{6.19}$$

The mass loss should decrease linearly with the energy received by the planet from stellar XEUV emission.

The mass loss can be described as a function of distance and the function $B(t)$, which takes the evolution of stellar luminosity into account.

$$\dot{m}(t) = -\frac{B(t)}{r^2}. \tag{6.20}$$

Finally one gets an equation for the variation of the semi-major axis a

$$\frac{da}{dt} = a\left(1 - 2\frac{a}{r}\right)\frac{\dot{\mu}}{\mu} + 2\frac{a^2 l_K \sin\theta}{\mu r^3}. \tag{6.21}$$

Let us assume an ejection speed of 100 km/s, a mass loss rate of $\dot{m} = -10^{15}$ g/s and $a = 0.05$ AU. We can make a simplification of the parameters in the formula above: $\mu = G(m(t) + m_*)$. The angle θ is defined by the geometry of mass ejection (Fig. 6.29).

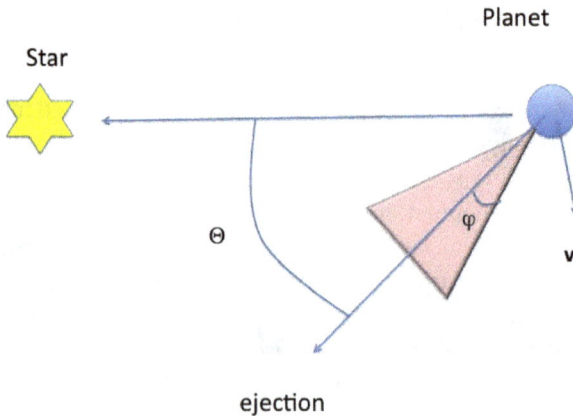

Fig. 6.29   Geometry of mass ejection according to Boué *et al.* [2012].

6.5.6 Case studies of hot Jupiters

In this section we summarize as an example the paper of Moses *et al.* [2011] where the atmospheres of the two exoplanets HD 189733b and HD 209458b are studied. The main characteristics of these two exoplanets are:

- HD 209458b
  - Star: Distance 47.1 pc, constellation Pegasus, mass 1.03 $M_\odot$, radius 1.14 $R_\odot$, temperature 6,000 K, age 4 ± 2 Gyr.
  - Orbital elements: Semi-major axis 0.045 AU, eccentricity 0.014, orbital period 3.52 d, inclination 86°.
  - Planet: Mass 0.71 M_J, radius 1.3 R_J, density 370 kg m^{-3}, surface gravity 0.9 g, temperature 1,140 K.

- HD 189733b
  - Star: Distance 19.5 pc, constellation Vulpecula, mass 0.84 $M_\odot$, radius 0.8 $R_\odot$, temperature 4,875 K, age 4.3 ± 2.8 Gyr.
  - Planet: Semi-major axis 0.030 AU, eccentricity 0.001, orbital period 2.218 d, inclination 86°, mass 1.16 M_J, radius 1.14 R_J, temperature 1,117 K.

The numerical code KINETICS, which was developed at Caltech/JPL, is a one-dimensional photochemistry/diffusion code. It is based on solving the following equation

$$\frac{\partial n_i}{\partial t} + \frac{\partial \Phi_i}{\partial z} = P_i - L_i, \qquad (6.22)$$

where i denotes the species, n_i the number density (cm^{-3}), Φ_i the flux (cm^{-2} s^{-1}), P_i the chemical production rate and L_i the loss rate, given in cm^{-3} s^{-1}; t denotes time and z the altitude.

For both planets, it can be shown that CH$_4$ and NH$_3$ are photochemically removed at higher altitudes. Disequilibrium chemistry also enhances atomic species, unsaturated hydrocarbons (particularly C$_2$H$_2$), some nitriles (particularly HCN) and radicals like OH, CH$_3$ and NH$_2$.

6.6 Exomoons

An extrasolar moon, or exomoon, is a natural satellite of an extrasolar planet. The Galilean satellites, especially Europa, are possible candidates for habitability because they contain a subsurface ocean heated by tidal interaction with Jupiter. Therefore, exomoons have to be considered as candidates for habitable objects and should also be considered here. A detailed review on exomoons can be found in Heller and Barnes [2013]. Potential exomoons are close to their host planets, the host star's activity plays only a minor role. The formation, habitability, and detection of extrasolar moons was reviewed in the work of Heller *et al.* [2014].

We give some basic information here.

6.6.1 *What affects habitability on an exomoon?*

Exomoons are affected by (i) their host planet and (ii) by the host star the planet is orbiting. Several special astrophysical effects have to be taken into account:

- There will be eclipses of the star by the planet.
- The moon's climate will be affected by two incoming fluxes: stellar flux and planetary flux.
- Tidal heating by the close host planet will be essential for life and habitability.
- The parent planet determines the moon's orbit and spin evolution.
- An orbit–spin resonance has to be expected. This is known for example in the case of the Earth and Moon where the resonance is 1:1, meaning that the Moon's revolution period around Earth equals its orbital spin period (rotation).

Considering exomoons in the inner circumstellar habitable zone, we can also expect, that the planet will become tidally locked, which means one hemisphere permanently faces the star. In the case of most exomoons, tides from the star are negligible because tidal forces decline following the relation

$$F_{\text{tidal}} \sim 1/r^3. \qquad (6.23)$$

Due to the larger distance r of the star from the planet–exomoon system, only the tidal forces of the planet are important. In most cases exomoons will be tidally locked to their planets rather than to the star. This means that the satellite's rotation period will be equal to its orbital period around the planet, a moon will orbit its planet in the planet's equatorial plane (Kozai mechanism) and the moon's rotational axis will be perpendicular to its orbital plane.

6.6.2 Orbital parameters of exomoons

In the restricted three-body problem, we consider the following masses: M_P planet, M_* host star and M_S satellite, where $M_S \ll M_P, M_*$. The satellite has no influence on the orbits of the two other objects. From the restricted three-body problem, the Hill sphere radius can be given as

$$R_H = a_p \left(\frac{M_p}{3M_*} \right)^{1/3}. \tag{6.24}$$

The critical semi-major axis of a satellite is the location of the outermost orbit at which the satellite still remains in the planet's gravitational field. This value is between $1/3$ and $1/2$ R_H. The Roche limit is given by

$$d_R = R_s \left(\frac{2M_P}{M_S} \right)^{1/3}. \tag{6.25}$$

A moon can only survive if it is far enough from the planet that it is outside the satellite-planet Roche limit and is close enough to the planet that it is within some fraction of the planet-star Hill radius. Theses two conditions give the following range for a_S [Scharf, 2006]

$$\left(\frac{3M_P}{2\pi\rho_S} \right)^{1/3} < a_S < 0.36(1 - e_P)a_P \left(\frac{M_P}{3M_*} \right)^{1/3}. \tag{6.26}$$

6.6.3 Atmospheres and magnetic field of exomoons

The stability of the atmosphere of exomoons depends on several factors:

- composition,
- intensity of the XUV and X-ray radiation from the host stars,
- surface gravity of the moon,
- magnetic field.

For example, nitrogen-dominated atmospheres (such as what is found on Saturn's moon Titan) can be stripped away by XUV radiation from young stars and also late-type stars because the upper atmosphere gets heated and expands and thermal escape occurs. Let us consider Saturn's largest satellite, Titan. It is at a distance of approximately 10 AU from the Sun. If we place Titan to a distance of only 1 AU from the Sun, the EUV radiation it would received would be 100 times more intense. That would lead to a rapid loss of its atmosphere because its surface gravity is lower than that of the Earth (Titan has a smaller mass than Earth). CO_2 provides a cooling of the atmosphere by IR radiation. Therefore, this mechanism could balance the thermal expansion.

A magnetic shield is also important for keeping an atmosphere stable. Studies have shown that a minimum mass is required for a magnetic field to provide long-term shielding. Here, long term means a timescale that is larger than 1 billion years. The minimum mass is about 0.1 Earth mass ($M_\oplus$). Tectonic activity can be expected for masses larger than $0.25\,M_\oplus$, and weak internal dynamos can be expected for masses $>0.25\,M_\oplus$. However, these limits can be considerably changed by other heat sources of exomoons such as tidal heating or radiogenic sources.

There is also an upper mass limit for the process of plate tectonics or dynamo action at about $2\,M_\oplus$. In that case, the pressure in the moon's interior increases as well as the mantle viscosity. This depresses the heat transfer from the core to the mantle. The dynamo becomes too weak to sustain a magnetic field and plate tectonics also disappear.

From these considerations the ideal habitable candidates for exomoons would be Earth-sized objects. In the solar system, the largest satellite is Ganymede. It has a radius of $0.4\,R_\oplus$ and a mass of $0.025\,M_\oplus$.

What do we know about the formation of exomoons? There are three possibilities:

(1) It has been shown that moons can be formed in the circumplanetary disk of giant planets having masses $\leq 10^{-4}$ times that of the planet's mass [Canup and Ward, 2006].
(2) Another way to explain large massive satellites around planets is capturing. Triton, a satellite of Neptune, is an example of a captured satellite.
(3) Exomoons could also have grown to larger masses by accretion processes.

F-type stars of spectral type between F5 V and F9.5 V, which possess Jupiter-type planets within or close to their climatological habitable zones, were investigated by Sato *et al.* [2017]. The Jupiter-type planets in these systems are in notably differing orbits with eccentricities between 0.08 (about Mars) and 0.72. The UV radiation environment on possible exomoons was also studied. The damage to DNA by UV radiation was found to be higher than damage to DNA on early Earth. Of course, this strongly depends on the orbital parameters of the planet.

6.6.4 *Radiation from plasma*

Planets may have a radiation belt that could be the source for ionizing radiation. Exomoons often orbit around their planets in a plasma environment that contains energetic particles. What are the effects of this radiation on the exomoon's atmosphere? Firstly, this ionizing radiation could strip away a moon's atmosphere, and, secondly, it could prevent the buildup of complex molecules on the moon's surface. Sputtering means that incident particles lose part of their energy to excitate target atoms and molecules in a planetary atmosphere to a higher energetic state in a planetary atmosphere.

Jupiter's Galilean satellites have been well studied for such processes. The main sources for sputtering on these satellites are ions like O^+, S^+, H^+. By exchanging energies, these heavy ions give rise to a flux of H_2O, O, O_2, H, H_2 from Ganymede's surface.

The H_2O and O_2 exospheres of Ganymede are studied in the paper by Plainaki *et al.* [2015]. The generation of Ganymede's exosphere is dominated by two processes: (i) surface release processes (sputtering, sublimation, radiolysis) and (ii) precipitation of the energetic ions from Jupiter's magnetosphere.

Therefore, habitability on an exomoon might be strongly decreased. An exomoon requires a power protection by its own magnetosphere. The magnetic field of Ganymede is about 750 nT. The habitability of the giant exomoons of Jupiter is discussed in connection with the JUICE mission [Grasset *et al.*, 2012]. JUICE is the acronym for JUpiter ICy moons Explorer, and this ESA mission is planned to be launched by 2022. It will perform detailed studies of the four Galilean moons. The main scientific objectives are to characterize the ocean layers, detect putative subsurface water reservoirs, study the chemistry essential to life on Europa and provide a subsurface sounding of that moon.

Theories for magnetic field generation on terrestrial planets show that the strength of this field depends on the planetary mass and the rotation frequency. The spin of exomoons is determined by the tides of their host planet. The rotation of an Earth-sized exomoon can be much faster than the the rotation of an Earth-sized planet orbiting a star, especially in the inner habitable zone. An exomoon must not be in tidal synchronization with the host star and the internal dynamo, and thus magnetic shielding will be enhanced. Some calculations show that even a strong convection in the planet's or moon's mantle could produce a considerable dynamo effect even in the case of a slowly rotating object.

Magnetic constraints on the habitability of exo-Earths and exomoons are discussed by Zuluaga [2013].

6.6.5 Tidal heating of exomoons

An important parameter is the tidal heating of a moon by its planet. Consider the case of the Moon. The obliquity is only 6.680 against its orbit. Tidal heating occurs mostly near the Moon's core. On Io, tidal heating mainly originates not because of an obliquity, but because of the high eccentricity of its orbit. Tidal core heating enhances the temperature gradient between core and mantle and, therefore, convection which could finally strengthen the magnetic field. In the case of mantle heating, the temperature gradient between the core and mantle becomes smaller and convection decreases.

In this context a Tidal Temperate Zone (TTZ) can be defined. This zone depends more on the orbital period of the moon and less on its radius. Tidal forces may be stronger in extrasolar systems because a large range of orbital parameters can be found. Using a so-called viscoelastic model, it has been shown that exomoons in the TTZ with orbital periods between 0.1 and 3.5 days could contain a melted subsurface liquid ocean. This strongly enhances the chances for habitability; the inner melting of the body moderates the surface temperature, acting like a thermostat [Dobos and Turner, 2015]. The inner ocean is protected against UV and X-rays as well as energetic particles by the overlying crust.

Consider a moon in a nonzero eccentricity orbit. Such an orbit can be maintained by orbital resonances with other moons (e.g. Io, Europa or Ganymede) or via resonances with the host planet orbit. $\dot{E}$ denotes the rate of tidal dissipation and H_T the surface heat flow:

$$H_T = \frac{\dot{E}}{4\pi R_S^2}. \qquad (6.27)$$

For H_T the following formula is found:

$$H_T = \frac{21}{38} \frac{G^{5/2}}{\mu Q} \left(\frac{3}{4\pi} \right)^{5/3} e_S^2 \frac{\rho_S^{1/3}}{a_S^{15/2}} M_S^{5/3} M_P^{5/2}, \qquad (6.28)$$

where μ is the satellite elastic rigidity, Q is the satellite specific dissipation function, R_S is the satellite radius and e_S the satellite orbital eccentricity.

6.6.6 *Hunt for exomoons*

So far (beginning of 2017), no confirmed exomoon detections were reported. One of the first interesting projects to begin the hunt for exomoons was the HEK (Hunt for Exomoons with Kepler) project [Kipping *et al.*, 2012]. The detection of moons around planets would complete our picture of planet/moon formation. Moreover, some of them could be habitable.

We have discussed the various techniques for detecting exoplanets; the detection of signatures of exomoons is even more difficult.

There are two broad categories of observational consequences of an exomoon in the transit lightcurve: dynamical variations and eclipse features. The method of transit timing variations, TTVs, works as follows: the motion of the planet around the planet–moon barycenter causes a planet to periodically transit early and late. There are also velocity-induced transit timing variations, TDV-V. The planet's velocity is inversely proportional to the duration of the transit. The transit timing variation is proportional to

$$\sim a_{\mathrm{SB}} M_s \tag{6.29}$$

where a_{SB} is the distance between the satellite and the planet body, M_S is the satellite's mass. This method favors the detection of moons at large separation; the TDV-V is proportional to

$$\sim a_{\mathrm{SB}}^{-1/2} M_S \tag{6.30}$$

which favors the detection of close moons. This is analogous to the radial velocity method of finding planets.

The second class of detection methods are eclipse methods. Signatures should be detectable in lightcurves during transit observations (see Fig. 6.30).

During its orbit, Io's ionosphere interacts with Jupiter's magnetosphere, to create a frictional current that causes radio wave emissions. These are called o-controlled decametric emissions. Similar emissions near known exoplanets could indicate where other moons exist. Another method is to measure the polarization of the emission of large planets. Large planets have extended atmospheres, and dust grains in their atmosphere produce a scattering and, therefore, a linear polarization. However, the disk-averaged polarization signal should be zero if the planet is spherically symmetric. If the planet rotates very fast, rotation-induced oblateness occurs, which causes a signal. When an exomoon transits a planetary atmosphere, such a

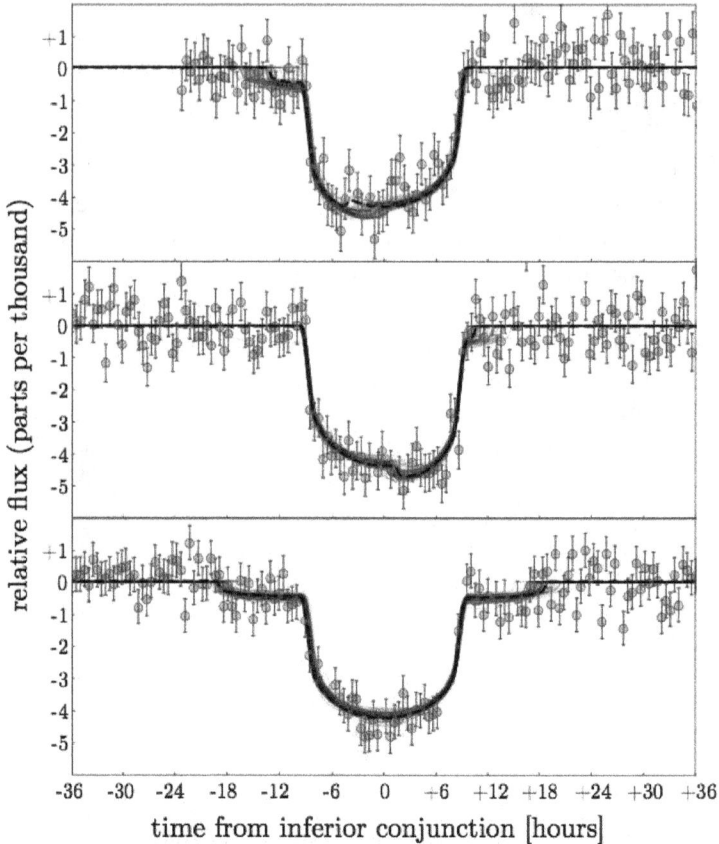

Fig. 6.30 Three transits of Kepler-1625b. The small dips in the lightcurve indicate the presence of an exomoon. Credit: Teachey/Kipping/Schmidt.

signal should also be measurable. In the IR, peak polarization signals due to the passage of an exomoon is expected between 0.1 and 0.3 [Sengupta and Marley, 2016].

6.7 Planets with no protection

In this section we study the case where objects have no protection against space weather effects from their hosting stars. Such planets are not interesting for life but they could have lost their shielding atmosphere or magnetosphere during their evolution and therefore they may have been habitable objects during their early stages.

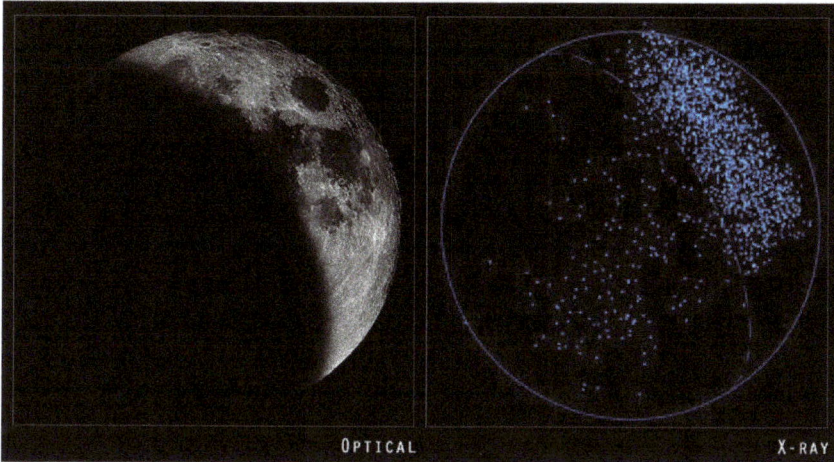

Fig. 6.31 Optical (left) and X-ray image of the Moon. The X-ray spots are produced by solar X-ray bombardment.

6.7.1 *Case study: Moon*

Let us consider space weather effects on the Moon. The Moon has no magnetosphere, and its atmosphere is negligible. Therefore, we can study the direct interaction between solar radiation, solar particles and the lunar surface. In Fig. 6.31, an optical image and an X-ray image of the Moon are shown. It is clearly seen that bright points appear in the X-ray image. These are caused by oxygen, magnesium, aluminum and silicon atoms. The X-rays are produced by fluorescence when solar X-rays bombard the Moon's surface, ejecting/exciting inner shell electrons.

Regolith covers almost the entire lunar surface; bedrock only protrudes on very steep-sided crater walls and the occasional lava channel. This regolith formed over the last 4.6 billion years from the impact of large and small meteoroids, from the steady bombardment of micrometeoroids and from solar and galactic charged particles breaking down surface rocks. The regolith is generally 4 to 5 m thick in mare areas and from 10 to 15 m in the older highland regions.

6.7.2 *Case study: Mercury and Asteroids*

Mercury's regolith, derived from the crustal bedrock, has been altered by a set of space weathering processes. An explanation and interpretation of its crustal composition is only reliable when we understand the nature of these processes. The case of Mercury is slightly different from that of the Moon since this planet has an exosphere. The processes that space weather the lunar surface are similar to those that form Mercury's exosphere (micrometeoroid flux and solar wind interactions) and are moderated by the local space environment and the presence of a global magnetic field. To comprehend how space weathering acts on Mercury's regolith, an understanding of how contributing processes act as an interactive system is needed.

Being the planet closest to the Sun, photon irradiation is up to an order of magnitude greater. This creates amorphous grain rims, chemically reducing the upper layers of grains to produce nanometer-scale particles of metallic iron, and depleting surface grains of volatile elements and alkali metals.

More details can be found, e.g. in the study of Domingue *et al.* [2014].

The bombardment by solar wind ions chemically reduces the top few atomic layers of the grain by sputtering away the oxygen molecules and leaving behind the iron that forms into small metallic particles. Bombardment of the surface by ultraviolet light can produce similar effects. Mercury has a global magnetic field, which can limit the ability of solar-wind ions to reach the surface, whereas ultraviolet light (photons) is not impeded by the magnetic field. In Fig. 6.32 these processes are illustrated.

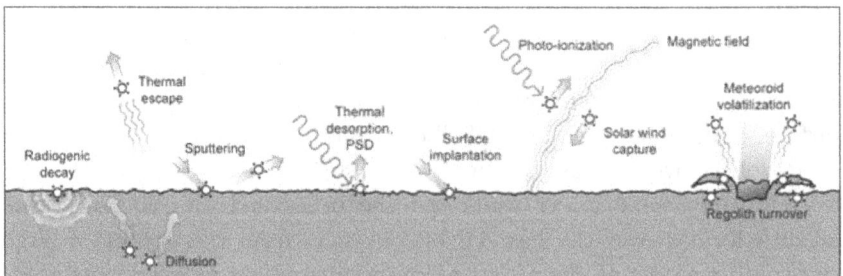

Fig. 6.32   Mercury: Space weathering effects on an atmosphereless object. Credit: Sam Savage.

Bibliography

Adibekyan, V., Delgado-Mena, E., Figueira, P., Sousa, S. G., Santos, N. C., Gonzalez Hernandez, J. I., Minchev, I., Faria, J. P., Israelian, G., Harutyunyan, G., Suarez-Andres, L., and Hakobyan, A. A. (2016). VizieR Online Data Catalog: Chemical abundances of solar analogues (Adibekyan+, 2016), *VizieR Online Data Catalog*, **359**.

Aitken, R. G. (1938). Is the solar system unique? *Leaflet of the Astronomical Society of the Pacific*, **3**, 98.

Anglada-Escudé, G., Arriagada, P., Vogt, S. S., Rivera, E. J., Butler, R. P., Crane, J. D., Shectman, S. A., Thompson, I. B., Minniti, D., Haghighipour, N., Carter, B. D., Tinney, C. G., Wittenmyer, R. A., Bailey, J. A., O'Toole, S. J., Jones, H. R. A., and Jenkins, J. S. (2012). A planetary system around the nearby M Dwarf GJ 667C with at least one super-earth in its habitable zone, *Astrophysical Journal Letters*, **751**, L16, doi:10.1088/2041-8205/751/1/L16, arXiv:1202.0446 [astro-ph.EP].

Antoniadou, K. I. (2016). Regular and chaotic orbits in the dynamics of exoplanets, *European Physical Journal Special Topics*, **225**, doi:10.1140/epjst/e2016-02651-6, arXiv:1604.04751 [astro-ph.EP].

Apai, D., Rackham, B. V., Lopez-Morales, M., Espinoza, N., Jordan, A., Osip, D., Lewis, N. K., Rodler, F., Fraine, J., Morley, C., Fortney, J. J., Bixel, A., ACCESS Team, and Earths in Other Solar Systems Team (2017). The ACCESS transiting exoplanets spectroscopy survey and the impact of heterogeneous stellar atmospheres on transit spectroscopy, in *American Astronomical Society Meeting Abstracts*, Vol. 229, p. 301.07.

Armitage, P. J. (2007). Lecture notes on the formation and early evolution of planetary systems, arXiv:astro-ph/0701485.

Arvidson, R. E. (2016). Aqueous history of Mars as inferred from landed mission measurements of rocks, soils, and water ice, *Journal of Geophysical Research (Planets)*, **121**, 1602–1626, doi:10.1002/2016JE005079.

Badman, S. V., Branduardi-Raymont, G., Galand, M., Hess, S. L. G., Krupp, N., Lamy, L., Melin, H., and Tao, C. (2015). Auroral processes at the giant planets: Energy deposition, emission mechanisms, morphology and spectra, *Space Science Review*, **187**, 99–179, doi:10.1007/s11214-014-0042-x.

Bailes, M., Bates, S. D., Bhalerao, V., Bhat, N. D. R., Burgay, M., Burke-Spolaor, S., D'Amico, N., Johnston, S., Keith, M. J., Kramer, M., Kulkarni, S. R., Levin, L., Lyne, A. G., Milia, S., Possenti, A., Spitler, L., Stappers, B., and van Straten, W. (2011). Transformation of a star into a planet in a millisecond pulsar binary, *Science*, **333**, 1717, doi:10.1126/science.1208890, arXiv:1108.5201 [astro-ph.SR].

Balona, L. A. (2015a). Flare stars across the H–R diagram, *IAU General Assembly*, **22**, 2242630.

Balona, L. A. (2015b). VizieR Online Data Catalog: Flare stars across the H–R diagram (Balona+, 2015), *VizieR Online Data Catalog*, **744**.

Barnes, R., Mullins, K., Goldblatt, C., Meadows, V. S., Kasting, J. F., and Heller, R. (2013). Tidal venuses: Triggering a climate catastrophe via tidal heating, *Astrobiology*, **13**, 225–250, doi:10.1089/ast.2012.0851, arXiv:1203.5104 [astro-ph.EP].

Barnes, S. A. (2003). On the rotational evolution of solar- and late-type stars, its magnetic origins, and the possibility of stellar gyrochronology, *Astrophysical Journal*, **586**, 464–479, doi:10.1086/367639, arXiv:astro-ph/0303631.

Baross, J. A. (1983). Growth of 'black smoker' bacteria at temperatures of at least 250°C, *Nature*, **303**, 423–426, doi:10.1038/303423a0.

Basu, S. (2016). Global seismology of the Sun, arXiv:1606.07071 [astro-ph.SR].

Batalha, N. M., Borucki, W. J., Bryson, S. T., Buchhave, L. A., Caldwell, D. A., Christensen-Dalsgaard, J., Ciardi, D., Dunham, E. W., Fressin, F., Gautier, T. N., III, Gilliland, R. L., Haas, M. R., Howell, S. B., Jenkins, J. M., Kjeldsen, H., Koch, D. G., Latham, D. W., Lissauer, J. J., Marcy, G. W., Rowe, J. F., Sasselov, D. D., Seager, S., Steffen, J. H., Torres, G., Basri, G. S., Brown, T. M., Charbonneau, D., Christiansen, J., Clarke, B., Cochran, W. D., Dupree, A., Fabrycky, D. C., Fischer, D., Ford, E. B., Fortney, J., Girouard, F. R., Holman, M. J., Johnson, J., Isaacson, H., Klaus, T. C., Machalek, P., Moorehead, A. V., Morehead, R. C., Ragozzine, D., Tenenbaum, P., Twicken, J., Quinn, S., VanCleve, J., Walkowicz, L. M., Welsh, W. F., Devore, E., and Gould, A. (2011). Kepler's first rocky planet: Kepler-10b, *Astrophysical Journal*, **729**, 27, doi: 10.1088/0004-637X/729/1/27, arXiv:1102.0605 [astro-ph.EP].

Batygin, K. and Laughlin, G. (2008). On the dynamical stability of the solar system, *Astrophysical Journal*, **683**, 1207–1216, doi:10.1086/589232, arXiv:0804.1946.

Beech, M. (2011). The past, present and future supernova threat to Earth's biosphere, *Astrophysics and Space Science*, **336**, 287–302, doi:10.1007/ s10509-011-0873-9.

Benedict, G. F., McArthur, B. E., Forveille, T., Delfosse, X., Nelan, E., Butler, R. P., Spiesman, W., Marcy, G., Goldman, B., Perrier, C., Jefferys, W. H., and Mayor, M. (2002). A mass for the extrasolar planet Gliese 876b determined from Hubble Space Telescope fine guidance sensor 3 astrometry and high-precision radial velocities, *Astrophysical Journal Letters*, **581**, L115–L118, doi:10.1086/346073, arXiv:astro-ph/0212101.

Benedict, G. F., McArthur, B. E., Gatewood, G., Nelan, E., Cochran, W. D., Hatzes, A., Endl, M., Wittenmyer, R., Baliunas, S. L., Walker, G. A. H., Yang, S., Kürster, M., Els, S., and Paulson, D. B. (2006). The extrasolar planet ε Eridani b: Orbit and mass, *Astronomical Journal*, **132**, 2206–2218, doi:10.1086/508323, arXiv:astro-ph/0610247.

Bennett, D. P. (2008). Detection of extrasolar planets by gravitational microlensing, in *Exoplanets: Detection, Formation, Properties, Habitability* (Springer), p. 47, doi:10.1007/978-3-540-74008-7_3.

Benz, A. O., Conway, J., and Gudel, M. (1998). First VLBI images of a main-sequence star, *Astronomy and Astrophysics*, **331**, 596–600.

Berdyugina, S. V. (2016). Polarized scattering and biosignatures in exoplanetary atmospheres, arXiv:1607.06874 [astro-ph.EP].

Bertucci, C., Duru, F., Edberg, N., Fraenz, M., Martinecz, C., Szego, K., and Vaisberg, O. (2011). The induced magnetospheres of Mars, Venus, and Titan, *Space Science Review*, **162**, 113–171, doi:10.1007/s11214-011-9845-1.

Beust, H., Bonneau, D., Mourard, D., Lafrasse, S., Mella, G., Duvert, G., and Chelli, A. (2011). On the use of the Virtual Observatory to select calibrators for phase-referenced astrometry of exoplanet–host stars, *Monthly Notices of Royal Astronomical Society*, **414**, 108–115, doi:10.1111/j.1365-2966.2011.18260.x.

Birn, J., Forbes, T. G., Hones, E. W., Jr., Bame, S. J., and Paschmann, G. (1981). On the velocity distribution of ion jets during substorm recovery, *Journal of Geophys. Research*, **86**, 9001–9006, doi:10.1029/JA086iA11p09001.

Boley, A. C., Payne, M. J., Corder, S., Dent, W. R. F., Ford, E. B., and Shabram, M. (2012). Constraining the planetary system of fomalhaut using high-resolution ALMA observations, *Astrophysical Journal Letters* **750**, L21, doi:10.1088/2041-8205/750/1/L21, arXiv:1204.0007 [astro-ph.EP].

Borovikov, S. N. and Pogorelov, N. V. (2014). Voyager 1 near the Heliopause, *Astrophysical Journal Letters*, **783**, L16, doi:10.1088/2041-8205/783/1/L16.

Boué, G., Figueira, P., Correia, A. C. M., and Santos, N. C. (2012). Orbital migration induced by anisotropic evaporation. Can hot Jupiters form hot Neptunes? *Astronomy and Astrophysics*, **537**, L3, doi:10.1051/0004-6361/201118084, arXiv:1109.2805 [astro-ph.EP].

Brunini, A. (1992). A theoretical dynamical limit of planet X's mass based on its perturbations on Uranus and Neptune, *Astronomy and Astrophysics* **265**, 324–327.

Canup, R. M. and Ward, W. R. (2006). A common mass scaling for satellite systems of gaseous planets, *Nature*, **441**, 834–839, doi:10.1038/nature04860.

Carroll, T. A., Kopf, M., Strassmeier, K. G., Ilyin, I., and Tuominen, I. (2009). Zeeman–Doppler imaging of II Peg, in K. G. Strassmeier, A. G. Kosovichev, and J. E. Beckman (eds.), *Cosmic Magnetic Fields: From*

Planets, to Stars and Galaxies, *IAU Symposium*, Vol. 259, pp. 437–438, doi:10.1017/S1743921309031044.

Carvalho, J. P. S., Mourão, D. C., de Moraes, R. V., Prado, A. F. B. A., and Winter, O. C. (2016). Exoplanets in binary star systems: On the switch from prograde to retrograde orbits, *Celestial Mechanics and Dynamical Astronomy*, **124**, 73–96, doi:10.1007/s10569-015-9650-3.

Cassan, A., Kubas, D., Beaulieu, J.-P., Dominik, M., Horne, K., Greenhill, J., Wambsganss, J., Menzies, J., Williams, A., Jørgensen, U. G., Udalski, A., Bennett, D. P., Albrow, M. D., Batista, V., Brillant, S., Caldwell, J. A. R., Cole, A., Coutures, C., Cook, K. H., Dieters, S., Prester, D. D., Donatowicz, J., Fouqué, P., Hill, K., Kains, N., Kane, S., Marquette, J.-B., Martin, R., Pollard, K. R., Sahu, K. C., Vinter, C., Warren, D., Watson, B., Zub, M., Sumi, T., Szymański, M. K., Kubiak, M., Poleski, R., Soszynski, I., Ulaczyk, K., Pietrzyński, G., and Wyrzykowski, L. (2012). One or more bound planets per Milky Way star from microlensing observations, *Nature*, **481**, 167–169, doi:10.1038/nature10684, arXiv:1202.0903 [astro-ph.EP].

Cayrel de Strobel, G., Knowles, N., Hernandez, G., and Bentolila, C. (1981). In search of real solar twins, *Astronomy and Astrophysics*, **94**, 1–11.

Chabrier, G. and Küker, M. (2006). Large-scale α^2-dynamo in low-mass stars and brown dwarfs, *Astronomy and Astrophysics* **446**, 1027–1037, doi:10.1051/0004-6361:20042475, astro-ph/0510075.

Charbonneau, D., Brown, T. M., Latham, D. W., and Mayor, M. (2000). Detection of planetary transits across a Sun-like star, *Astrophysical Journal Letters*, **529**, L45–L48, doi:10.1086/312457, astro-ph/9911436.

Charbonneau, P. (2016). Solar physics: Dynamo theory questioned, *Nature*, **535**, 500–501, doi:10.1038/535500a.

Chauvin, G., Lagrange, A.-M., Zuckerman, B., Dumas, C., Mouillet, D., Song, I., Beuzit, J.-L., Lowrance, P., and Bessell, M. S. (2005). A companion to AB Pic at the planet/brown dwarf boundary, *Astronomy and Astrophysics*, **438**, L29–L32, doi:10.1051/0004-6361:200500111, arXiv:astro-ph/0504658.

Chiavassa, A., Caldas, A., Selsis, F., Leconte, J., Von Paris, P., Bordé, P., Magic, Z., Collet, R., and Asplund, M. (2017). Measuring stellar granulation during planet transits, *Astronomy and Astrophysics* **597**, A94, doi:10.1051/0004-6361/201528018, arXiv:1609.08966 [astro-ph.EP].

Collier, C. A. (2005). Mapping of stellar surfaces with Doppler and Zeeman Doppler imaging, *Highlights of Astronomy*, **13**, 805.

Cooper, G., Horz, F., O'Leary, A., and Chang, S. (2013). The impact and oxidation survival of selected meteoritic compounds: Signatures of asteroid organic material on planetary surfaces, in *Lunar and Planetary Science Conference*, *Lunar and Planetary Science Conference*, Vol. 44, p. 1868.

Costa, A. D., Canto Martins, B. L., Bravo, J. P., Paz-Chinchón, F., das Chagas, M. L., Leão, I. C., Pereira de Oliveira, G., Rodrigues da Silva, R., Roque, S., de Oliveira, L. L. A., Freire da Silva, D., and De Medeiros, J. R. (2015). Kepler rapidly rotating giant stars, *Astrophysical Journal Letters*, **807**, L21, doi:10.1088/2041-8205/807/2/L21, arXiv:1506.06644 [astro-ph.SR].

Cuntz, M., von Bloh, W., Schröder, K.-P., Bounama, C., and Franck, S. (2012). Habitability of super-Earth planets around main-sequence stars including red giant branch evolution: models based on the integrated system approach, *International Journal of Astrobiology*, **11**, 15–23, doi: 10.1017/S1473550411000280, arXiv:1107.5714 [astro-ph.EP].

Davenport, J. R. A. (2016). The Kepler catalog of stellar flares, *Astrophysical Journal*, **829**, 23, doi:10.3847/0004-637X/829/1/23, arXiv:1607.03494 [astro-ph.SR].

de Mooij, E. J. W., Brogi, M., de Kok, R. J., Koppenhoefer, J., Nefs, S. V., Snellen, I. A. G., Greiner, J., Hanse, J., Heinsbroek, R. C., Lee, C. H., and van der Werf, P. P. (2012). Optical to near-infrared transit observations of super-Earth GJ 1214b: Water-world or mini-Neptune? *Astronomy and Astrophysics*, **538**, A46, doi:10.1051/0004-6361/201117205, arXiv:1111.2628 [astro-ph.EP].

del Burgo, C. and Allende Prieto, C. (2017). Accurate parameters for HD 209458 and its planet from HST spectrophotometry, arXiv:1703.01449 [astro-ph.SR].

Deming, L. D. and Seager, S. (2017). Illusion and reality in the atmospheres of exoplanets, *Journal of Geophysical Research (Planets)*, **122**, 53–75, doi: 10.1002/2016JE005155.

Dobos, V. and Turner, E. L. (2015). Viscoelastic models of tidally heated exomoons, *Astrophysical Journal*, **804**, 41, doi:10.1088/0004-637X/804/1/41, arXiv:1502.07090 [astro-ph.EP].

Domingue, D. L., Chapman, C. R., Killen, R. M., Zurbuchen, T. H., Gilbert, J. A., Sarantos, M., Benna, M., Slavin, J. A., Schriver, D., Trávníček, P. M., Orlando, T. M., Sprague, A. L., Blewett, D. T., Gillis-Davis, J. J., Feldman, W. C., Lawrence, D. J., Ho, G. C., Ebel, D. S., Nittler, L. R., Vilas, F., Pieters, C. M., Solomon, S. C., Johnson, C. L., Winslow, R. M., Helbert, J., Peplowski, P. N., Weider, S. Z., Mouawad, N., Izenberg, N. R., and McClintock, W. E. (2014). Mercury's Weather-Beaten surface: Understanding mercury in the context of lunar and asteroidal space weathering studies, *Space Science Review*, **181**, 121–214, doi:10.1007/s11214-014-0039-5.

Dominik, M., Albrow, M. D., Beaulieu, J.-P., Caldwell, J. A. R., DePoy, D. L., Gaudi, B. S., Gould, A., Greenhill, J., Hill, K., Kane, S., Martin, R., Menzies, J., Naber, R. M., Pel, J.-W., Pogge, R. W., Pollard, K. R., Sackett, P. D., Sahu, K. C., Vermaak, P., Watson, R., and Williams, A. (2002). The PLANET microlensing follow-up network: Results and prospects for the detection of extra-solar planets, *Planetary and Space Science*, **50**, 299–307, doi:10.1016/S0032-0633(01)00126-X.

Egeland, R., Soon, W. H., Baliunas, S. L., Hall, J. C., Pevtsov, A. A., and Henry, G. W. (2016). Dynamo sensitivity in solar analogs with 50 years of Ca II H & K Activity, in *AAS/Solar Physics Division Meeting*, *AAS/Solar Physics Division Meeting*, Vol. 47, pp. 203–207.

Ehrenreich, D., Bourrier, V., Wheatley, P. J., Lecavelier des Etangs, A., Hébrard, G., Udry, S., Bonfils, X., Delfosse, X., Désert, J.-M., Sing, D. K., and Vidal-Madjar, A. (2015). A giant comet-like cloud of hydrogen escaping the warm Neptune-mass exoplanet GJ 436b, *Nature*, **522**, 459–461, doi: 10.1038/nature14501, arXiv:1506.07541 [astro-ph.EP].

Esteves, L. J., de Mooij, E. J. W., Jayawardhana, R., Watson, C., and de Kok, R. (2017). A search for water in a super-earth atmosphere: High-resolution optical spectroscopy of 55Cancri e, *Astronomical Journal* **153**, 268, doi: 10.3847/1538-3881/aa7133, arXiv:1705.03022 [astro-ph.EP].

Estrela, R. and Valio, A. (2017). Superflare UV flashes impact on Kepler-96 system: A glimpse of habitability when the ozone layer first formed on Earth, arXiv:1708.05400 [astro-ph.EP].

Feulner, G. (2012). The faint young Sun problem, *Reviews of Geophysics*, **50**, RG2006, doi:10.1029/2011RG000375, arXiv:1204.4449 [astro-ph.EP].

Fressin, F., Torres, G., Charbonneau, D., Bryson, S. T., Christiansen, J., Dressing, C. D., Jenkins, J. M., Walkowicz, L. M., and Batalha, N. M. (2013). The false positive rate of Kepler and the occurrence of planets, *Astrophysical Journal*, **766**, 81, doi:10.1088/0004-637X/766/2/81, arXiv:1301.0842 [astro-ph.EP].

Fressin, F., Torres, G., Rowe, J. F., Charbonneau, D., Rogers, L. A., Ballard, S., Batalha, N. M., Borucki, W. J., Bryson, S. T., Buchhave, L. A., Ciardi, D. R., Désert, J.-M., Dressing, C. D., Fabrycky, D. C., Ford, E. B., Gautier, T. N., III, Henze, C. E., Holman, M. J., Howard, A., Howell, S. B., Jenkins, J. M., Koch, D. G., Latham, D. W., Lissauer, J. J., Marcy, G. W., Quinn, S. N., Ragozzine, D., Sasselov, D. D., Seager, S., Barclay, T., Mullally, F., Seader, S. E., Still, M., Twicken, J. D., Thompson, S. E., and Uddin, K. (2012). Two Earth-sized planets orbiting Kepler-20, *Nature*, **482**, 195–198, doi:10.1038/nature10780, arXiv:1112.4550 [astro-ph.EP].

Funk, B., Pilat-Lohinger, E., and Eggl, S. (2015). Can there be additional rocky planets in the Habitable zone of tight binary stars with a known gas giant? *Monthly Notices of Royal Astronomical Society*, **448**, 3797–3805, doi:10. 1093/mnras/stv253, arXiv:1505.07069 [astro-ph.EP].

Gao, P., Marley, M. S., Zahnle, K., Robinson, T. D., and Lewis, N. K. (2017). Sulfur hazes in giant exoplanet atmospheres: Impacts on reflected light spectra, arXiv:1701.00318 [astro-ph.EP].

Gliese, W. and Jahreiss, H. (2015). VizieR Online Data Catalog: Catalogue of nearby stars, Revised version 2 (Gliese+ 1979), *VizieR Online Data Catalog*, **5035**.

Goździewski, K., Maciejewski, A. J., and Migaszewski, C. (2007). On the extrasolar multiplanet system around HD 160691, *Astrophysical Journal*, **657**, 546–558, doi:10.1086/510554, astro-ph/0608279.

Grasset, O., Prieto-Ballesteros, O., Titov, D., Erd, C., Bunce, E., Coustenis, A., Blanc, M., Coates, A., Fletcher, L., van Hoolst, T., Hussmann, H., Jaumann, R., Krupp, N., Tortora, P., Tosi, F., and Wielders, A. (2012). Habitability of the giant icy moons: Current knowledge and future insights from the JUICE mission, in *European Planetary Science Congress 2012*, EPSC2012-925.

Guinan, E. F., Engle, S. G., and Durbin, A. (2016). Living with a red dwarf: Rotation and X-ray and ultraviolet properties of the halo population

Kapteyn's star, *Astrophysical Journal*, **821**, 81, doi:10.3847/0004-637X/821/2/81, arXiv:1602.01912 [astro-ph.SR].

Guo, J., Lin, L., Bai, C., and Liu, J. (2017). The effects of the Reimers η on the solar rotational period when our Sun evolves to the RGB tip, *Astrophysics and Space Science*, **362**, 15, doi:10.1007/s10509-016-2978-7.

Hackman, T., Lehtinen, J., Rosén, L., Kochukhov, O., and Käpylä, M. J. (2016). Zeeman–Doppler imaging of active young solar-type stars, *Astronomy and Astrophysics*, **587**, A28, doi:10.1051/0004-6361/201527320, arXiv:1509.02285 [astro-ph.SR].

Haghighipour, N. and Kaltenegger, L. (2013). Calculating the habitable zone of binary star systems. II. P-type binaries, *Astrophysical Journal*, **777**, 166, doi:10.1088/0004-637X/777/2/166, arXiv:1306.2890 [astro-ph.EP].

Haynes, K., Mandell, A., and Deming, D. (2014). Exoplanet secondary eclipses using WFC3, in *American Astronomical Society Meeting Abstracts*, Vol. 223, p. 230.08.

Heller, R. and Barnes, R. (2013). Exomoon habitability constrained by illumination and tidal heating, *Astrobiology*, **13**, 18–46, doi:10.1089/ast.2012.0859, arXiv:1209.5323 [astro-ph.EP].

Heller, R., Williams, D., Kipping, D., Limbach, M. A., Turner, E., Greenberg, R., Sasaki, T., Bolmont, É., Grasset, O., Lewis, K., Barnes, R., and Zuluaga, J. I. (2014). Formation, habitability, and detection of extrasolar moons, *Astrobiology*, **14**, 798–835, doi:10.1089/ast.2014.1147, arXiv:1408.6164 [astro-ph.EP].

Hempelmann, A., Mittag, M., Gonzalez-Perez, J. N., Schmitt, J. H. M. M., Schröder, K. P., and Rauw, G. (2016). Measuring rotation periods of solar-like stars using TIGRE. A study of periodic CaII H+K S-index variability, *Astronomy and Astrophysics*, **586**, A14, doi:10.1051/0004-6361/201526972.

Heng, K. (2017). *Exoplanetary Atmospheres: Theoretical Concepts and Foundations* (Princeton University Press).

Heng, K., Menou, K., and Phillipps, P. J. (2011). Atmospheric circulation of tidally locked exoplanets: A suite of benchmark tests for dynamical solvers, *Monthly Notices of Royal Astronomical Society*, **413**, 2380–2402, doi:10.1111/j.1365-2966.2011.18315.x, arXiv:1010.1257 [astro-ph.EP].

Howard, A. W., Sanchis-Ojeda, R., Marcy, G. W., Johnson, J. A., Winn, J. N., Isaacson, H., Fischer, D. A., Fulton, B. J., Sinukoff, E., and Fortney, J. J. (2013). A rocky composition for an Earth-sized exoplanet, *Nature*, **503**, 381–384, doi:10.1038/nature12767, arXiv:1310.7988 [astro-ph.EP].

Huitson, C. M., Sing, D. K., Vidal-Madjar, A., Ballester, G. E., Lecavelier des Etangs, A., Désert, J.-M., and Pont, F. (2012). Temperature–Pressure profile of the hot Jupiter HD 189733b from HST sodium observations: Detection of upper atmospheric heating, arXiv:1202.4721 [astro-ph.EP].

Hulot, G., Finlay, C. C., Constable, C. G., Olsen, N., and Mandea, M. (2010). The magnetic field of planet earth, *Space Science Review*, **152**, 159–222, doi:10.1007/s11214-010-9644-0.

Hyodo, R., Charnoz, S., Ohtsuki, K., and Genda, H. (2017). Ring formation around giant planets by tidal disruption of a single passing large Kuiper Belt object, *Icarus*, **282**, 195–213, doi:10.1016/j.icarus.2016.09.012, arXiv:1609.02396 [astro-ph.EP].

Iorio, L. (2010). Orbital effects of Sun's mass loss and the Earth's fate, *Natural Science*, **2**, 329–337, doi:10.4236/ns.2010.24041, gr-qc/0511138.

Jain, S., Stewart, I. F., Schneider, N. M., Deighan, J., Stiepen, A., Evans, J. S., Stevens, M. H., Chaffin, M., Crismani, M. M. J., McClintock, B., Montmessin, F., Thiemann, E., Epavier, F., Chamberlin, P. C., and Jakosky, B. M. (2016). Martian upper atmosphere response to solar EUV flux and soft X-ray flare, *AGU Fall Meeting Abstracts*.

Johnson, J. L. and Li, H. (2012). The first planets: The critical metallicity for planet formation, *Astrophysical Journal*, **751**, 81, doi:10.1088/0004-637X/751/2/81, arXiv:1203.4817 [astro-ph.EP].

Kaltenegger, L. and Haghighipour, N. (2013). Calculating the habitable zone of binary star systems. I. S-type binaries, *Astrophysical Journal*, **777**, 165, doi:10.1088/0004-637X/777/2/165, arXiv:1306.2889 [astro-ph.EP].

Kane, S. R. and Hinkel, N. R. (2013). The habitable zones of circumbinary planetary systems, in *American Astronomical Society Meeting Abstracts*, Vol. 221, pp. 343–329.

Kasting, J. F., Whitmire, D. P., and Reynolds, R. T. (1993). Habitable zones around main sequence stars, *Icarus*, **101**, 108–128, doi:10.1006/icar.1993.1010.

Kelley, D. S., Baross, J. A., and Delaney, J. R. (2002). Volcanoes, fluids, and life at mid-ocean ridge spreading centers, *Annual Review of Earth and Planetary Sciences*, **30**, 385–491, doi:10.1146/annurev.earth.30.091201.141331.

Khodachenko, M. L., Ribas, I., Lammer, H., Grießmeier, J.-M., Leitner, M., Selsis, F., Eiroa, C., Hanslmeier, A., Biernat, H. K., Farrugia, C. J., and Rucker, H. O. (2007). Coronal mass ejection (CME) activity of low mass m stars as an important factor for the habitability of terrestrial exoplanets. I. CME impact on expected magnetospheres of earth-like exoplanets in close-in habitable zones, *Astrobiology*, **7**, 167–184, doi:10.1089/ast.2006.0127.

Khodachenko, M. L., Sasunov, Y., Arkhypov, O. V., Alexeev, I. I., Belenkaya, E. S., Lammer, H., Kislyakova, K. G., Odert, P., Leitzinger, M., and Güdel, M. (2014). Stellar CME activity and its possible influence on exoplanets' environments: Importance of magnetospheric protection, in B. Schmieder, J.-M. Malherbe, and S. T. Wu (eds.), *Nature of Prominences and their Role in Space Weather*, *IAU Symposium*, Vol. 300, pp. 335–346, doi:10.1017/S1743921313011174.

Kiang, N. Y., Segura, A., Tinetti, G., Govindjee, Blankenship, R. E., Cohen, M., Siefert, J., Crisp, D., and Meadows, V. S. (2007). Spectral signatures of photosynthesis. II. Coevolution with other stars and the atmosphere on extrasolar worlds, *Astrobiology*, **7**, 252–274, doi:10.1089/ast.2006.0108, astro-ph/0701391.

Kipping, D. M., Bakos, G. Á., Buchhave, L., Nesvorný, D., and Schmitt, A. (2012). The hunt for exomoons with Kepler (HEK). I. Description of a new observational project, *Astrophysical Journal*, **750**, 115, doi:10.1088/0004-637X/750/2/115, arXiv:1201.0752 [astro-ph.EP].

Kitzmann, D. (2017). Clouds in the atmospheres of extrasolar planets. V. The impact of CO_2 ice clouds on the outer boundary of the habitable zone, *Astronomy and Astrophysics*, **600**, A111, doi:10.1051/0004-6361/201630029, arXiv:1701.07513 [astro-ph.EP].

Kochukhov, O. (2016). Doppler and Zeeman Doppler Imaging of Stars, in J.-P. Rozelot and C. Neiner (eds.), *Lecture Notes in Physics, Berlin Springer Verlag*, Vol. 914, p. 177, doi:10.1007/978-3-319-24151-7_9.

Kowalski, A. F., Hawley, S. L., Holtzman, J. A., Wisniewski, J. P., and Hilton, E. J. (2010). A white light megaflare on the dM4.5e star YZ CMi, *Astrophysical Journal Letters*, **714**, L98–L102, doi:10.1088/2041-8205/714/1/L98, arXiv:1003.3057 [astro-ph.SR].

Kulikov, Y. N., Lammer, H., Lichtenegger, H. I. M., Terada, N., Ribas, I., Kolb, C., Langmayr, D., Lundin, R., Guinan, E. F., Barabash, S., and Biernat, H. K. (2006). Atmospheric and water loss from early Venus, *Planetary and Space Science*, **54**, 1425–1444, doi:10.1016/j.pss.2006.04.021.

Kurokawa, H. and Nakamoto, T. (2014). Mass-loss evolution of close-in exoplanets: Evaporation of hot jupiters and the effect on population, *Astrophysical Journal*, **783**, 54, doi:10.1088/0004-637X/783/1/54, arXiv:1401.2511 [astro-ph.EP].

Lammer, H., Bredehöft, J. H., Coustenis, A., Khodachenko, M. L., Kaltenegger, L., Grasset, O., Prieur, D., Raulin, F., Ehrenfreund, P., Yamauchi, M., Wahlund, J.-E., Grießmeier, J.-M., Stangl, G., Cockell, C. S., Kulikov, Y. N., Grenfell, J. L., and Rauer, H. (2009). What makes a planet habitable? *The Astronomy and Astrophysics Review*, **17**, 181–249, doi:10.1007/s00159-009-0019-z.

Lammer, H., Chassefière, E., Karatekin, Ö., Morschhauser, A., Niles, P. B., Mousis, O., Odert, P., Möstl, U. V., Breuer, D., Dehant, V., Grott, M., Gröller, H., Hauber, E., and Pham, L. B. S. (2013). Outgassing history and escape of the martian atmosphere and water inventory, *Space Science Review*, **174**, 113–154, doi:10.1007/s11214-012-9943-8, arXiv:1506.06569 [astro-ph.EP].

Lin, H. W. and Loeb, A. (2015). Statistical signatures of panspermia in exoplanet surveys, *Astrophysical Journal Letters*, **810**, L3, doi:10.1088/2041-8205/810/1/L3, arXiv:1507.05614 [astro-ph.EP].

Lineweaver, C. (2007). The galactic habitable zone and the age distribution of complex life in the Milky Way, in *American Astronomical Society Meeting Abstracts*, Vol. 210; *Bulletin of the American Astronomical Society*, Vol. 39, p. 173.

Lineweaver, C. H., Fenner, Y., and Gibson, B. K. (2004). The galactic habitable zone and the age distribution of complex life in the Milky Way, *Science*, **303**, 59–62, doi:10.1126/science.1092322, astro-ph/0401024.

Lingam, M. (2016). Analytical approaches to modelling panspermia — beyond the mean-field paradigm, *Monthly Notices of Royal Astronomical Society*, **455**, 2792–2803, doi:10.1093/mnras/stv2533.

Luu, J. X. and Jewitt, D. C. (2002). Kuiper Belt objects: Relics from the accretion disk of the Sun, *Annual Review of Astronomy and Astrophysics*, **40**, 63–101, doi:10.1146/annurev.astro.40.060401.093818.

Maeder, A. and Meynet, G. (2012). Rotating massive stars: From first stars to gamma ray bursts, *Reviews of Modern Physics*, **84**, 25–63, doi:10.1103/ RevModPhys.84.25.

Maehara, H., Shibayama, T., Notsu, S., Notsu, Y., Nagao, T., Kusaba, S., Honda, S., Nogami, D., and Shibata, K. (2012). Superflares on solar-type stars, *Nature*, **485**, 478–481, doi:10.1038/nature11063.

Malik, M., Grosheintz, L., Mendonça, J. M., Grimm, S. L., Lavie, B., Kitzmann, D., Tsai, S.-M., Burrows, A., Kreidberg, L., Bedell, M., Bean, J. L., Stevenson, K. B., and Heng, K. (2017). HELIOS: An open-source, GPU-accelerated radiative transfer code for self-consistent exoplanetary atmospheres, *Astronomical Journal*, **153**, 56, doi:10.3847/1538-3881/153/2/56, arXiv:1606.05474 [astro-ph.EP].

Marois, C., Zuckerman, B., Konopacky, Q. M., Macintosh, B., and Barman, T. (2010). Images of a fourth planet orbiting HR 8799, *Nature*, **468**, 1080–1083, doi:10.1038/nature09684, arXiv:1011.4918 [astro-ph.EP].

Marsden, S. C., Petit, P., Jeffers, S. V., Morin, J., Fares, R., Reiners, A., do Nascimento, J.-D., Aurière, M., Bouvier, J., Carter, B. D., Catala, C., Dintrans, B., Donati, J.-F., Gastine, T., Jardine, M., Konstantinova-Antova, R., Lanoux, J., Lignières, F., Morgenthaler, A., Ramìrez-Vèlez, J. C., Théado, S., Van Grootel, V., and BCool Collaboration (2014). A BCool magnetic snapshot survey of solar-type stars, *Monthly Notices of Royal Astronomical Society*, **444**, 3517–3536, doi:10.1093/mnras/stu1663, arXiv:1311.3374 [astro-ph.SR].

Marsh, T. R. and Horne, K. (1988). Images of accretion discs. II — Doppler tomography, *Monthly Notices of Royal Astronomical Society*, **235**, 269–286, doi:10.1093/mnras/235.1.269.

Massol, H., Hamano, K., Tian, F., Ikoma, M., Abe, Y., Chassefière, E., Davaille, A., Genda, H., Güdel, M., Hori, Y., Leblanc, F., Marcq, E., Sarda, P., Shematovich, V. I., Stökl, A., and Lammer, H. (2016). Formation and evolution of protoatmospheres, *Space Science Review*, doi:10.1007/ s11214-016-0280-1.

Matsakos, T., Uribe, A., and Konigl, A. (2015). 3D modelling of magnetized star-planet interactions: Cometary-type tails and in-spiraling flows, in G. T. van Belle and H. C. Harris (eds.), *18th Cambridge Workshop on Cool Stars, Stellar Systems, and the Sun*, pp. 373–376.

Mayor, M. and Queloz, D. (1995). A Jupiter-mass companion to a solar-type star, *Nature*, **378**, 355–359, doi:10.1038/378355a0.

Meftah, M., Hauchecorne, A., Irbah, A., and Bush, R. I. (2015). On solar oblateness measurements during the current solar cycle 24, *AGU Fall Meeting Abstracts*.

Meibom, S., Barnes, S. A., Platais, I., Gilliland, R. L., Latham, D. W., and Mathieu, R. D. (2015). A spin-down clock for cool stars from observations of a

2.5-billion-year-old cluster, *Nature*, **517**, 589–591, doi:10.1038/nature14118, arXiv:1501.05651 [astro-ph.SR].

Meléndez, J., Ramírez, I., Karakas, A. I., Yong, D., Monroe, T. R., Bedell, M., Bergemann, M., Asplund, M., Tucci Maia, M., Bean, J., do Nascimento, J.-D., Jr., Bazot, M., Alves-Brito, A., Freitas, F. C., and Castro, M. (2014). 18 Sco: A solar twin rich in refractory and neutron-capture elements. *Implications for Chemical Tagging*, *Astrophysical Journal*, **791**, 14, doi: 10.1088/0004-637X/791/1/14, arXiv:1406.5244 [astro-ph.SR].

Miller, S. L. (1953). A production of amino acids under possible primitive earth conditions, *Science*, **117**, 528–529, doi:10.1126/science.117.3046.528.

Moschou, S.-P., Drake, J. J., Cohen, O., Alvarado-Gomez, J. D., and Garraffo, C. (2017). A monster CME obscuring a demon star flare, arXiv:1710.07361 [astro-ph.SR].

Moses, J. I., Fouchet, T., Bézard, B., Lellouch, E., Gladstone, G. R., Feuchtgruber, H., and Allen, M. (2001). Comparative planetology: Lessons from photochemical modeling of the upper atmospheres of Jupiter and Saturn, in *Bulletin of the American Astronomical Society*, Vol. 33, p. 1044.

Moses, J. I., Marley, M. S., Zahnle, K., Line, M. R., Fortney, J. J., Barman, T. S., Visscher, C., Lewis, N. K., and Wolff, M. J. (2016). On the composition of young, directly imaged giant planets, *Astrophysical Journal*, **829**, 66, doi: 10.3847/0004-637X/829/2/66, arXiv:1608.08643 [astro-ph.EP].

Moses, J. I., Visscher, C., Fortney, J. J., Showman, A. P., Lewis, N. K., Griffith, C. A., Klippenstein, S. J., Shabram, M., Friedson, A. J., Marley, M. S., and Freedman, R. S. (2011). Disequilibrium Carbon, Oxygen, and Nitrogen Chemistry in the atmospheres of HD 189733b and HD 209458b, *Astrophysical Journal*, **737**, 15, doi:10.1088/0004-637X/737/1/15, arXiv:1102.0063 [astro-ph.EP].

Nordhaus, J. and Spiegel, D. S. (2013). On the orbits of low-mass companions to white dwarfs and the fates of the known exoplanets, *Monthly Notices of Royal Astronomical Society*, **432**, 500–505, doi:10.1093/mnras/stt569, arXiv:1211.1013 [astro-ph.SR].

Noyes, R. W., Hartmann, L. W., Baliunas, S. L., Duncan, D. K., and Vaughan, A. H. (1984). Rotation, convection, and magnetic activity in lower main-sequence stars, *Astrophysical Journal*, **279**, 763–777, doi:10.1086/161945.

O'Brien, D. P. and Sykes, M. V. (2011). The origin and evolution of the asteroid belt implications for Vesta and Ceres, *Space Science Review*, **163**, 41–61, doi:10.1007/s11214-011-9808-6.

Odert, P., Leitzinger, M., Hanslmeier, A., and Lammer, H. (2017). Stellar coronal mass ejections — I. Estimating occurrence frequencies and mass-loss rates, *Monthly Notices of Royal Astronomical Society*, **472**, 876–890, doi:10.1093/mnras/stx1969, arXiv:1707.02165 [astro-ph.SR].

Odrzywolek, A. and Rafelski, J. (2016). Classification of exoplanets according to density, arXiv:1612.03556 [astro-ph.EP].

Oishi, M., Watanabe, K., and Kamaya, H. (2017). Expansion of habitable zone around low mass stars by the supply of UV radiation from CME, *Memoirs of the National Defence Academy of Japan*, Vol. 55, pp. 9–16.

Oreshenko, M., Heng, K., and Demory, B.-O. (2016). Optical phase curves as diagnostics for aerosol composition in exoplanetary atmospheres, *Monthly Notices of Royal Astronomical Society*, **457**, 3420–3429, doi:10.1093/mnras/stw133, arXiv:1601.03050 [astro-ph.EP].

Pallavicini, R., Golub, L., Rosner, R., Vaiana, G. S., Ayres, T., and Linsky, J. L. (1981). Relations among stellar X-ray emission observed from Einstein, stellar rotation and bolometric luminosity, *Astrophysical Journal*, **248**, 279–290, doi:10.1086/159152.

Parker, E. N. (1970). The origin of solar magnetic fields, *Annual Review of Astronomy and Astrophysics*, **8**, 1, doi:10.1146/annurev.aa.08.090170.000245.

Parker, E. N. (1977). The generation of magnetic fields in astrophysical bodies. XI — The effect of magnetic buoyancy on the growth and migration of dynamo waves in the sun, *Astrophysical Journal*, **215**, 370–373, doi:10.1086/155366.

Parker, E. N. (1979). *Cosmical Magnetic Fields: Their Origin and their Activity* (Oxford University Press).

Parker, E. T., Cleaves, H. J., Callahan, M. P., Dworkin, J. P., Glavin, D. P., Lazcano, A., and Bada, J. L. (2011). Prebiotic synthesis of methionine and other sulfur-containing organic compounds on the primitive Earth: A contemporary reassessment based on an unpublished 1958 Stanley Miller Experiment, *Origins of Life and Evolution of the Biosphere*, **41**, 201–212, doi:10.1007/s11084-010-9228-8.

Pierrehumbert, R. T. (2010). *Principles of Planetary Climate* (Cambridge University Press).

Pillitteri, I., Wolk, S. J., Lopez-Santiago, J., and Sciortino, S. (2015). The (phased?) activity of stars hosting hot jupiters, in G. T. van Belle and H. C. Harris (eds.), *18th Cambridge Workshop on Cool Stars, Stellar Systems, and the Sun*, Vol. 18, pp. 583–588.

Pinsonneault, M. H., Matt, S., and MacGregor, K. B. (2013). Angular momentum and mass loss from magnetized solar-like winds, in *American Astronomical Society Meeting Abstracts*, Vol. 221, pp. 252–206.

Plainaki, C., Lilensten, J., Radioti, A., Andriopoulou, M., Milillo, A., Nordheim, T. A., Dandouras, I., Coustenis, A., Grassi, D., Mangano, V., Massetti, S., Orsini, S., and Lucchetti, A. (2016). Planetary space weather: Scientific aspects and future perspectives, *Journal of Space Weather and Space Climate*, **6**, 27, A31, doi:10.1051/swsc/2016024.

Plainaki, C., Milillo, A., Massetti, S., Mura, A., Jia, X., Orsini, S., Mangano, V., De Angelis, E., and Rispoli, R. (2015). The H_2O and O_2 exospheres of Ganymede: The result of a complex interaction between the jovian magnetospheric ions and the icy moon, *Icarus*, **245**, 306–319, doi:10.1016/j.icarus.2014.09.018.

Porto de Mello, G. F. and da Silva, L. (1997). HR 6060: The closest ever solar twin? *Astrophysical Journal Letters*, **482**, L89, doi:10.1086/310693.

Prantzos, N. (2008). On the "Galactic Habitable Zone", *Space Science Review*, **135**, 313–322, doi:10.1007/s11214-007-9236-9, astro-ph/0612316.

Pye, J. P., Rosen, S., Fyfe, D., and Schröder, A. C. (2015). A survey of stellar X-ray flares from the XMM-Newton serendipitous source catalogue: HIPPARCOS-Tycho cool stars, *Astronomy and Astrophysics*, **581**, A28, doi:10.1051/0004-6361/201526217, arXiv:1506.05289 [astro-ph.SR].

Ramirez, R. M. and Kaltenegger, L. (2017). A volcanic hydrogen habitable zone, *Astrophysical Journal Letters*, **837**, L4, doi:10.3847/2041-8213/aa60c8, arXiv:1702.08618 [astro-ph.EP].

Rasio, F. A., Tout, C. A., Lubow, S. H., and Livio, M. (1996). Tidal decay of close planetary orbits, *Astrophysical Journal*, **470**, 1187, doi:10.1086/177941, astro-ph/9605059.

Reimers, D. (1975). Circumstellar absorption lines and mass loss from red giants, *Memoires of the Societe Royale des Sciences de Liege*, **8**, 369–382.

Reynolds, R. T., McKay, C. P., and Kasting, J. F. (1987). Europa, tidally heated oceans, and habitable zones around giant planets, *Advances in Space Research*, **7**, 125–132, doi:10.1016/0273-1177(87)90364-4.

Rogers, L. A., Bodenheimer, P., Lissauer, J. J., and Seager, S. (2011). Formation and structure of low-density exo-Neptunes, *Astrophysical Journal*, **738**, 59, doi:10.1088/0004-637X/738/1/59, arXiv:1106.2807 [astro-ph.EP].

Rugheimer, S., Segura, A., Kaltenegger, L., and Sasselov, D. (2015). UV surface environment of Earth-like planets orbiting FGKM stars through geological evolution, *Astrophysical Journal*, **806**, 137, doi:10.1088/0004-637X/806/1/137, arXiv:1506.07200 [astro-ph.EP].

Sato, S., Wang, Z., and Cuntz, M. (2017). Climatological and ultraviolet-based habitability of possible exomoons in F-star systems, *Astronomische Nachrichten*, **338**, 413–427, doi:10.1002/asna.201613279, arXiv:1503.02560 [astro-ph.SR].

Savanov, I. S. and Dmitrienko, E. S. (2015). Activity and cool spots on the surfaces of G-type stars with superflares from observations with the Kepler Space Telescope, *Astronomy Reports*, **59**, 879–887, doi:10.1134/S1063772915090073.

Scharf, C. A. (2006). The potential for tidally heated icy and Temperate Moons around exoplanets, *Astrophysical Journal*, **648**, 1196–1205, doi:10.1086/505256, astro-ph/0604413.

Schmitt, J. H. M. M., Schröder, K.-P., Rauw, G., Hempelmann, A., Mittag, M., González-Pérez, J. N., Czesla, S., Wolter, U., Jack, D., Eenens, P., and Trinidad, M. A. (2014). TIGRE: A new robotic spectroscopy telescope at Guanajuato, Mexico, *Astronomische Nachrichten*, **335**, 787, doi:10.1002/asna.201412116.

Schubert, G. and Soderlund, K. M. (2011). Planetary magnetic fields: Observations and models, *Physics of the Earth and Planetary Interiors*, **187**, 92–108, doi:10.1016/j.pepi.2011.05.013.

Schwarzschild, M. (1948). On noise arising from the solar granulations, *Astrophysical Journal*, **107**, 1, doi:10.1086/144983.

Sears, D. W. G. (2015). The explored asteroids: Science and exploration in the space age, *Space Science Review*, **194**, 139–235, doi:10.1007/s11214-015-0202-7.

Selsis, F., Kasting, J. F., Levrard, B., Paillet, J., Ribas, I., and Delfosse, X. (2007). Habitable planets around the star Gliese 581? *Astronomy and Astrophysics*, **476**, 1373–1387, doi:10.1051/0004-6361:20078091, arXiv:0710.5294.

Sengupta, S. and Marley, M. S. (2016). Detecting exomoons around self-luminous giant exoplanets through polarization, *Astrophysical Journal*, **824**, 76, doi: 10.3847/0004-637X/824/2/76, arXiv:1604.04773 [astro-ph.SR].

Shannon, A., Wu, Y., and Lithwick, Y. (2016). Forming the cold classical Kuiper Belt in a light disk, *Astrophysical Journal*, **818**, 175, doi:10.3847/0004-637X/818/2/175, arXiv:1510.01323 [astro-ph.EP].

Simon, T., Ayres, T. R., Redfield, S., and Linsky, J. L. (2002). Limits on chromospheres and convection among the main-sequence a stars, *Astrophysical Journal*, **579**, 800–809, doi:10.1086/342941.

Singer, K. N., McKinnon, W. B., Pappalardo, R. T., and Khurana, K. K. (2009). Europa: Perspectives on an ocean world, *AGU Fall Meeting Abstracts*.

Skumanich, A. (1972). Time scales for CA II emission decay, rotational braking, and lithium depletion, *Astrophysical Journal*, **171**, 565, doi:10.1086/151310.

Snell, C., Gomez Leal, I., Kaltenegger, L., and Jennings, R. (2017). How obliquitiy influences the climate of aquaplanets, in *American Astronomical Society Meeting Abstracts*, Vol. 229, p. 245.20.

Solanki, S. K. and Krivova, N. A. (2011). Analyzing solar cycles, *Science*, **334**, 916, doi:10.1126/science.1212555.

Southam, G., Rothschild, L. J., and Westall, F. (2007). The geology and habitability of terrestrial planets: Fundamental requirements for life, *Space Science Review*, **129**, 7–34, doi:10.1007/s11214-007-9148-8.

Southworth, J., Mancini, L., Madhusudhan, N., Molliere, P., Ciceri, S., and Henning, T. (2016). Detection of the atmosphere of the 1.6 Earth mass exoplanet GJ 1132b, arXiv:1612.02425 [astro-ph.EP].

Spitoni, E., Matteucci, F., and Sozzetti, A. (2014). The galactic habitable zone of the Milky Way and M31 from chemical evolution models with gas radial flows, *Monthly Notices of Royal Astronomical Society*, **440**, 2588–2598, doi: 10.1093/mnras/stu484, arXiv:1403.2268.

Sundin, M. (2006). The galactic habitable zone in barred galaxies, *International Journal of Astrobiology*, **5**, 325–326, doi:10.1017/S1473550406003065.

Svensmark, H. (2012). Evidence of nearby supernovae affecting life on Earth, *Monthly Notices of Royal Astronomical Society*, **423**, 1234–1253, doi:10.1111/j.1365-2966.2012.20953.x, arXiv:1210.2963 [astro-ph.SR].

Tinetti, G., Liang, M., Beaulieu, J., Yung, Y. L., Carey, S., Ribas, I., Tennyson, J., Barber, B., Allard, N., Ballester, G., Sing, D., and Selsis, F. (2007). Water vapour in the atmosphere of an extrasolar planet, in *Bulletin of the American Astronomical Society*, Vol. 38, p. 467.

Tourpali, K., Schuurmans, C. J. E., van Dorland, R., Steil, B., and Brühl, C. (2003). Stratospheric and tropospheric response to enhanced solar UV radiation: A model study, *Geophysical Research Letter*, **30**, 1231, doi: 10.1029/2002GL016650.

Triaud, A. (2016). Exoplanets: Migration of giants, *Nature*, **537**, 496–497, doi: 10.1038/nature19430.

Valle, G., Dell'Omodarme, M., Prada Moroni, P. G., and Degl'Innocenti, S. (2014). Evolution of the habitable zone of low-mass stars. Detailed stellar models and analytical relationships for different masses and chemical compositions, *Astronomy and Astrophysics*, **567**, A133, doi:10.1051/0004-6361/201323350, arXiv:1405.7486 [astro-ph.SR].

Vallée, J. P. (2011). Magnetic fields in the nearby Universe, as observed in solar and planetary realms, stars, and interstellar starforming nurseries, *New Astronomy Reviews*, **55**, 23–90, doi:10.1016/j.newar.2011.01.001.

Villadsen, J., Hallinan, G., and Bourke, S. (2016). Radio spectroscopy of stellar flares: Magnetic reconnection & CME shocks in stellar coronae, in A. G. Kosovichev, S. L. Hawley, and P. Heinzel (eds.), *Solar and Stellar Flares and their Effects on Planets, IAU Symposium*, Vol. 320, pp. 191–195, doi: 10.1017/S1743921316008644.

Vogt, S. S. and Penrod, G. D. (1983). Doppler Imaging of spotted stars — application to the RS Canum Venaticorum star HR 1099, *Publ. Astr. Soc. of the Pacific*, **95**, 565–576, doi:10.1086/131208.

Von Damm, K. L. (1990). Seafloor hydrothermal activity: Black smoker chemistry and chimneys, *Annual Review of Earth and Planetary Sciences*, **18**, 173, doi:10.1146/annurev.ea.18.050190.001133.

von Steiger, R. and Zurbuchen, T. H. (2016). Solar metallicity derived from *in situ* solar wind composition, *Astrophysical Journal*, **816**, 13, doi:10.3847/0004-637X/816/1/13.

Wächtershäuser, G. (1989). The case for an autotrophic origin, *Origins of Life and Evolution of the Biosphere*, **19**, 423–424, doi:10.1007/BF02388930.

Wächtershäuser, G. (1990). The case for the chemoautotrophic origin of life in an iron-sulfur world, *Origins of Life and Evolution of the Biosphere*, **20**, 173–176, doi:10.1007/BF01808279.

Walkowicz, L. M., Basri, G., Batalha, N., Gilliland, R. L., Jenkins, J., Borucki, W. J., Koch, D., Caldwell, D., Dupree, A. K., Latham, D. W., Meibom, S., Howell, S., Brown, T. M., and Bryson, S. (2011). White-light flares on cool stars in the Kepler quarter 1 data, *Astronomical Journal*, **141**, 50, doi:10.1088/0004-6256/141/2/50, arXiv:1008.0853 [astro-ph.SR].

Wallner, A., Feige, J., Kinoshita, N., Paul, M., Fifield, L. K., Golser, R., Honda, M., Linnemann, U., Matsuzaki, H., Merchel, S., Rugel, G., Tims, S. G., Steier, P., Yamagata, T., and Winkler, S. R. (2016). Recent near-Earth supernovae probed by global deposition of interstellar radioactive ^{60}Fe, *Nature*, **532**, 69–72, doi:10.1038/nature17196.

Walsh, R. W. (2001). Solar Magnetohydrodynamics, in A. Hanslmeier, M. Messerotti, and A. Veronig (eds.), *Astrophysics and Space Science Library*, Vol. 259, p. 129, doi:10.1007/978-94-010-0760-3_5.

Wolszczan, A. and Kuchner, M. (2010). *Planets Around Pulsars and other Evolved Stars: The Fates of Planetary Systems* (University of Arizona Press), pp. 175–190.

Wood, B. E., Müller, H.-R., Redfield, S., and Edelman, E. (2014). Evidence for a weak wind from the young sun, *Astrophysical Journal Letters*, **781**, L33, doi:10.1088/2041-8205/781/2/L33.

Woods, T. N., Eparvier, F. G., Fontenla, J., Harder, J., Kopp, G., McClintock, W. E., Rottman, G., Smiley, B., and Snow, M. (2004). Solar irradiance variability during the October 2003 solar storm period, *Geophysical Research Letter*, **31**, L10802, doi:10.1029/2004GL019571.

Wright, N. J., Drake, J. J., Mamajek, E. E., and Henry, G. W. (2011). The stellar–activity–rotation relationship and the evolution of stellar dynamos, *Astrophysical Journal*, **743**, 48, doi:10.1088/0004-637X/743/1/48, arXiv:1109.4634 [astro-ph.SR].

Zahnle, K. (2008). Cosmic impacts and climates of the terrestrial planets, *AGU Fall Meeting Abstracts*.

Zuluaga, J. I. (2013). Magnetic constraints on the habitability of exoearths and exomoons, *AGU Spring Meeting Abstracts*.

Zuluaga, J. I. and Cuartas, P. A. (2012). The role of rotation in the evolution of dynamo-generated magnetic fields in Super Earths, *Icarus*, **217**, 88–102, doi:10.1016/j.icarus.2011.10.014, arXiv:1101.0691 [astro-ph.EP].

Index

www.ingramcontent.com/pod-product-compliance
Lightning Source LLC
Chambersburg PA
CBHW050552190326
41458CB00007B/2010